S0-AUX-856

Roberts BOOK STORE
CASH FOR BOOKS
ANY TIME . . .
USED PRICE

Solutions/Resource Manual to Accompany

The World of the Cell

SECOND EDITION

Wayne M. Becker
The University of Wisconsin-Madison

The Benjamin/Cummings Publishing Company, Inc.
Redwood City, California • Menlo Park, California
Reading, Massachusetts • New York • Don Mills, Ontario • Wokingham
Amsterdam • Bonn • Sydney • Singapore • Tokyo • Madrid • San Juan

Sponsoring Editor: David Rogelberg
Production Coordinator: Alyssa Weiner
Composition: Ideas to Images
Cover Design: Victoria Ann Philp

Copyright © 1991 by The Benjamin/Cummings Publishing Company, Inc.

All rights reserved. No part of this publication may be reproduced, stored in a retrieval system, or transmitted, in any form or by any means, electronic, mechanical, photocopying, recording, or otherwise, without the prior written permission of the publisher. Printed in the United States of America. Published simultaneously in Canada.

ISBN 0-8053-0871-7

4 5 6 7 8 9 10-CRS-99 98 97 96 95 94

The Benjamin/Cummings Publishing Company, Inc.
390 Bridge Parkway
Redwood City, California 94065

Contents

Preface

This Solutions Manual includes complete answers to all problems at the end of each chapter in the second edition of *The World of the Cell* (The Benjamin/Cummings Publishing Company, Inc., 1991). The inclusion of problem sets in the text reflects my conviction that one learns science not just by reading or hearing about it, but by working with it. The availability of detailed solutions is intended both to confirm understanding where answers are correct and to provide guidance in the appropriate problem-solving skills where they are not. At the instructor's discretion, this manual can either be made available to students through the local bookstore or be used by the instructor as a resource for homework and exam questions.

I am deeply indebted to David W. Deamer, co-author of the above text, and to Peter B. Armstrong, Joel W. Goodman, David N. Gunn, and Jeanette E. Natzle, for providing answers to the problems that they wrote for the second edition. I nonetheless retain responsibility for any errors of omission or commission that are to be found here, and I welcome feedback from users of this manual. Please address comments and suggestions directly to me.

Wayne M. Becker
Department of Botany
B115 Birge Hall
University of Wisconsin
Madison, Wisconsin 53706

1

The World of the Cell: A Preview

1. (a) C   (d) B   (g) G   (j) G

   (b) B   (e) G   (h) B, C   (k) C

   (c) G   (f) C   (i) B   (l) B

2. Both the ultracentrifuge and the electron microscope proved to be especially powerful new tools for exploring the structure and function of cells and especially of organelles. The electron microscope allowed cellular structure to be examined in much greater detail and led to the detection of structures such as ribosomes, microtubules, and microfilaments, which are not even visible with the light microscope. The ultracentrifuge permitted isolation of subcellular fractions, thereby facilitating biochemical investigation of their specific functions. A comparable series of events both chronologically and in terms of its impact on experimentation and understanding of genetics might be the elucidation of the structure and function of DNA and the "cracking" of the genetic code.

3. (a) Bacterial cell: $V = \pi r^2 h = (3.14)(0.5)^2(2.0) = 1.57 \ \mu m^3$.

   Liver cell: $V = 4\pi r^3/3 = 4(3.14)(10)^3/3 = 4200 \ \mu m^3$.

   Palisade cell: $V = \pi r^2 h = (3.14)(10)^2(35) = 11,000 \ \mu m^3$.

   (b) Bacterial cells in a liver cell: $4200/1.57 = 2700$.

   (c) Liver cells in a palisade cell: $11,000/4200 = 2.63$.

4. (a) Light microscope: limit of resolution = 200 nm. Since 1 membrane has a thickness of 8 nm, the number of membranes that must be aligned laterally is 200/8 = **25 membranes.**

   Electron microscope: limit of resolution = 0.1–0.2 nm. With a thickness of 8 nm, a single membrane can be seen using an electron microscope. (To contrast light and electron microscopy in terms of resolving power, compare parts (a) and (b) of Figure 1-3, looking specifically at the nuclear membranes.)

   (b) Liver cell: V = 4200 μm^3 (from Problem 3a).

   Ribosome: V = $4\pi r^3/3$ = $4(3.14)(0.0125)^3/3$ = 8.2 x 10^{-6} μm^3.

   Ribosomes in a liver cell: 4200/(8.2 x 10^{-6}) = **5 x 10^8.**

   (c) Bacterial cell: V = 1.57 μm^3 (from Problem 3a).

   DNA molecule: V = $\pi r^2 h$ = $(3.14)(1 \text{ nm})^2(1.36 \text{ mm})$

   = $(3.14)(0.001 \ \mu m)^2(1.36 \times 10^3 \ \mu m)$

   = $(3.14)(1 \times 10^{-6} \ \mu m^2)(1.36 \times 10^3 \ \mu m)$

   = $(3.14)(1.36)(10^{-3})$ μm^3 = 4.3 x 10^{-3} μm^3.

   DNA = (4.3 x 10^{-3})/1.57 = 0.0027 = **0.27% of cell volume.**

5. (a) Initially thought to be true because animal cells do not have cell walls, which made it hard to distinguish individual cells using the crude microscopes available to early investigators; shown by Schwann to be incorrect for cartilage cells, which have well defined boundaries of collagen fibers, and later extended to all animal cells.

   (b) Initially thought to be true because living organisms seem to increase in complexity spontaneously, unlike other systems known to early chemists or physicists; misconception laid to rest by Wohler's demonstration that urea, a compound made by living organisms, could be synthesized in the laboratory from an inorganic starting compound.

(c) Originally thought to be true because the order of nucleotide monomers in DNA was erroneously considered to be an invariant tetranucleotide repeating sequence; disproved by Avery et al. for bacteria (1944) and by Hershey and Chase for bacterial viruses (1952).

(d) Initially thought to be true because of the demonstration by Pasteur that yeast cells were needed for alcoholic fermentation; Buchner showed later (1897) that extracts from yeast cells could substitute for intact cells, an effect we now know to be due to the presence in the extracts of the enzymes that catalyze the various reactions in the fermentation process.

6. (a) Consistent with early electron microscopy in which membranes appeared as two parallel electron-dense lines (thought to be the outer protein layer) separated by an unstained space (thought to be the lipid interior of the membrane); disproved by demonstration that membrane proteins are globular structures located within the membrane, not just on its surface.

(b) Postulated because of the early demonstration that high-energy intermediates are involved in ATP generation during the glycolytic pathway; now known not to be the case for ATP generation by mitochondria, where the driving force turns out to be the energy of an electrochemical proton gradient, with no high-energy phosphorylated intermediates involved.

(c) Shown to be true for the photosynthetic algae used for the first studies of ^{14}C fixation; now understood to be true only for so-called "C_3 plants," since "C_4 plants" are known in which the initial product of carbon fixation is a four-carbon compound instead.

(d) Believed to be universally true until the recent discovery of parasites called trypanosomes in which the enzymes of the glycolytic pathway responsible for the conversion of sugar into pyruvate are compartmentalized into a membrane-bounded organelle called the *glycosome*.

(e)   Thought to be universally true until the recent discovery of Z-DNA in which the two strands form a left-handed helix.

(f)   Thought to be universally true until the recent discovery that the genetic code used in mitochondria is different in several aspects from the code that appears to operate everywhere else.

2
The Chemistry of the Cell

1. (a) Carbon is a smaller atom than silicon (atomic weight of 12 instead of 28). It therefore forms especially stable covalent bonds, since the strength of a covalent bond is inversely proportional to the atomic weights of the elements involved.

    (b) Carbon combines with oxygen to form simple gaseous molecules and is therefore not removed from circulation in an aerobic environment, as occurs with silicon.

    (c) Carbon is one of only three elements (the other two are oxygen and nitrogen) that readily form strong multiple bonds. Carbon forms both double and triple bonds; the former especially are important in biological compounds, allowing the resonance structures of aromatic compounds.

    (d) Carbon also has four valence electrons, but it is much less abundant in the earth's crust than silicon (0.19% versus 28% of the earth's crust); carbon is not obviously fitter in this regard.

    (e) Polymers of carbon are stable in water because of the strength of the carbon-carbon bond (see part a). They are also readily soluble if they have hydrophilic groups. Stability and solubility are important biologically because most of life involves an aqueous milieu.

2. (a) T   (c) F   (e) T

    (b) T   (d) T   (f) X

3. (a) E = 28,600/λ; λ = 28,600/E = 28,600/83 = **344 nm.**

    (b) Less stable; its cutoff comes at 409 nm.

    (c) More stable; it can only be broken by ultraviolet light with wavelengths less than 196 nm.

4. (a) Four (all combinations of alternate configurations at carbon atoms 2 and 3).

    (b) Eight (all combinations of alternate configurations at carbon atoms 2, 3, and 4).

    (c) No stereoisomerism.

    (d) Four (all combinations of alternate configurations at both asymmetric carbon atoms).

5. (a) Hexadecane lacks a hydrophilic group and is therefore insoluble in an aqueous milieu.

    (b) Compound ii; compound i is a hydrocarbon and would not be soluble in an aqueous milieu. (Compound ii is the amino acid isoleucine.)

6. It is this asymmetry that renders water a polar molecule; most of the desirable properties of water as a solvent depend on its polarity.

7. (a) A single enzyme or set of enzymes can be used to add each successive monomeric unit.

    (b) Water is a readily available reactant in an essentially aqueous world.

8. (a) TMV virions self-assemble spontaneously without the input of energy or information, which means that all of the information necessary to direct their assembly must be already present in the RNA and/or proteins.

   (b) The strain-specific assembly of TMV is determined by the RNA, not the coat protein.

   (c) The information necessary to direct self-assembly of TMV virions appears to reside in the coat protein monomers.

   (d) The self-assembly of TMV virions is specific for TMV RNA.

   (e) The most stable configuration for TMV virions is achieved by the 3:1 ratio of nucleotides and coat protein monomers, and is therefore the product formed upon self-assembly regardless of the starting ratio of nucleotides and monomers.

9. The question is: "What are the advantages of subunit assembly in the elaboration of biological macromolecules and cellular structure?"

3

The Macromolecules
of the Cell

1. (a) 2, 4, 8   (d) 1, 4, 11

   (b) 5, 9, 11, 12   (e) 5, 9, 11, 12

   (c) 5, 6, 11, 12   (f) 5, 6, 11, 12

2.

Bond	Amino Acids	Levels of Structure
Peptide (covalent)	All	Primary
Hydrogen	All	Secondary
Hydrophobic	Leucine	Tertiary, Quaternary
Ionic	Glutamate	Tertiary, Quaternary
Sulfhydryl	Cysteine	Tertiary, Quaternary

3. (a) Interior: valine, phenylalanine (hydrophobic).

   Exterior: aspartate, lysine (charged; hydrophilic).

   Either: glycine, alanine (small R groups; no strong affinity for or aversion to water).

   (b) Alanine; phenylalanine; glutamate; methionine (the less polar member of each pair).

   (c) The free sulfhydryl group is polar and ionizable; the disulfide bond is much less polar.

4. The hydrophobic "patches" on the surfaces of the α and β subunits of hemoglobin represent sites of interaction between the subunits and as such are not on the surface of the intact tetrameric hemoglobin molecule, even though they are on the surface of the unassembled subunits.

5. (a) The amino acid glutamate is hydrophilic and ionizes at cellular pH, whereas valine is hydrophobic and nonionic. Substitution of the latter for the former is likely to change the chemical nature of that part of the molecule significantly.

    (b) Aspartate, cysteine.

    (c) Yes, if the substitutions are always of like-for-like amino acids in terms of chemical properties.

6. (a) To pull on both ends of an α-keratin polypeptide is to pull against the hydrogen bonds that account for its helical structure; these "give" readily, allowing the polypeptide to be stretched to its full, uncoiled length, at which point you would begin to pull against the covalent peptide bonds. For fibroin, you are pulling against the covalent peptide bonds immediately. (As an analogy, you might compare pulling on opposite ends of a coiled spring versus a straight length of uncoiled wire.)

    (b) Fibroin consists mainly of the two smallest amino acids, so it has few bulky R groups and can accommodate the constraints of a pleated sheet. Keratin, on the other hand, has most of the amino acids present, and the distance between bulky R groups is maximized when these protrude from a twisted helical shape.

7. (a) Hair proteins are first treated with a sulfhydryl reducing agent to break disulfide bonds and thereby destroy much of the natural tertiary structure and shape of the hair. After being "set" in the

desired shape, the hair is treated with an oxidizing agent to allow disulfide bonds to re-form, but now between different cysteine groups, as determined by the positioning imposed by the curlers. These unnatural disulfide bonds then stabilize the desired configuration.

(b) There are two reasons for the lack of permanence: (l) Disulfide bonds occasionally break and re-form spontaneously, allowing the hair proteins to return gradually to their original, thermo-dynamically more favorable shape. (2) Hair continues to grow, and the new α-keratin molecules will have the natural (correct) disulfide bonds.

(c) There is probably a genetic difference in the positioning of cysteine groups and hence in the formation of disulfide bonds.

8. (a) k, f, d   (d) none   (g) k, d

    (b) k   (e) k, f   (h) k, f, d

    (c) f, d   (f) f, d   (i) f, d

9. (a) Circumference $= \dfrac{4 \times 10^6 \text{ pairs}}{\text{molecule}} \times \dfrac{1 \text{ turn}}{10 \text{ pairs}} \times \dfrac{3.4 \text{ nm}}{\text{turn}}$

$$= 1.36 \times 10^6 \text{ nm} = \textbf{1.36 mm.}$$

This poses a packaging problem, since the circumference of the DNA is about 680 times the length of the whole cell!

(b) Circumference $= \pi d = (3.14)(1.8 \ \mu m) = 5.65 \ \mu m = \textbf{5650 nm.}$

$$\frac{5650 \text{ nm}}{\text{molecule}} \times \frac{1 \text{ turn}}{3.4 \text{ nm}} \times \frac{10 \text{ pairs}}{\text{turn}} = \textbf{16,600 nucleotide pairs/molecule.}$$

$$16,600 \text{ pairs} \times \frac{1 \text{ polypeptide}}{1500 \text{ pairs}} = \textbf{11 polypeptides.}$$

(c) $\dfrac{2.8 \times 10^9 \text{ g}}{\text{mole}} \times \dfrac{1 \text{ mole}}{6.02 \times 10^{23} \text{ molecules}} \times \dfrac{1 \text{ molecule}}{4 \times 10^6 \text{ pair}}$

$$= 1.16 \times 10^{-21} \text{ g/pair.}$$

$\dfrac{6.0 \times 10^{-12} \text{ g}}{\text{nucleus}} \times \dfrac{1 \text{ pair}}{1.16 \times 10^{-21} \text{ g}} = 5.17 \times 10^9 \text{ pairs per nucleus.}$

(d) $\dfrac{5.17 \times 10^9 \text{ pairs}}{\text{nucleus}} \times \dfrac{1 \text{ turn}}{10 \text{ pairs}} \times \dfrac{3.4 \text{ nm}}{\text{turn}} = 17.6 \times 10^8 \text{ nm}$

$$= 1.76 \text{ meters of DNA.}$$

10. (a) Compared to a linear molecule, a branched-chain polymer has more termini for addition or hydrolysis of glucose units per unit volume of polymer, thereby facilitating both the deposition and mobilization of glucose by providing more sites for enzymatic activity.

(b) Every branch point will have an $\alpha(1{\rightarrow}6)$ glycosidic bond that will have to be hydrolyzed. This is handled by the presence of an additional enzyme specific for the $\alpha(1{\rightarrow}6)$ bond.

(c) Endolytic cleavage breaks the molecule internally, creating additional ends for exolytic attack and thereby allowing the mobilization of more glucose per unit time.

(d) Cellulose molecules are rigid, linear rods which aggregate laterally into microfibrils. Branches in the molecule would generate side chains which would almost certainly make it difficult to pack the cellulose molecules into microfibrils, thereby decreasing the rigidity and strength of the microfibrils.

11.

Inulin

12. Phosphatidyl choline > fatty acid > triglyceride > cholesterol

[Two charged groups > carboxyl group > ester bonds > alcohol group]

13. The presence of an oleate will introduce a bend in one of the side chains that will closely approximate the shape of sphingomyelin, since the latter also has a double bond and hence a bend.

14. (a) Partial hydrogenation results in the partial reduction of C=C double bonds to C–C single bonds in the fatty acyl chains of the triglycerides.

(b) Before partial hydrogenation, the shortening was vegetable oil and therefore liquid at room temperature.

(c) Partial hydrogenation makes the shortening solid instead of liquid and allows it to be used as a substitute for animal fat.

4

Cells and Organelles

1. (a) T; the volume of a eukaryotic cell is usually several orders of magnitude greater than that of a prokaryotic cell.

   (b) T; some egg cells, neurons, and algal cells are examples.

   (c) F; a plasma membrane is common to all cells.

   (d) T; the ratio decreases with increasing cell size.

   (e) T; the ribosomes found in eukaryotic organelles are very similar to those present in prokaryotic cells.

   (f) F; these functions also occur in prokaryotic cells, but without compartmentalization into organelles.

2. (a) 25 mm (1 in.); large marble.

   (b) 25×1000 mm (1×40 in.); short fishing pole.

   (c) 7×200 mm ($\frac{1}{4} \times 8$ in.); drinking straw.

   (d) 0.5 m (20 in.); large beach ball.

   (e) 1×2 m ($3\frac{1}{4} \times 6\frac{1}{2}$ ft.); household hot water heater.

   (f) 2×8 m ($6\frac{1}{2} \times 26$ ft.); large gas storage cylinder.

   (g) 6 m (20 ft.); hot air balloon.

   (h) 20 m (65 ft.); municipal water tower tank.

   (i) 40×60 km (25×40 mi.); area of a large city.

(j)   1800 km (1200 mi.); distance from Minneapolis to New York City.

3.   (a)   Secretion; pancreatic cells synthesize a variety of digestive enzymes and hormones, which are then secreted into the intestinal tract (enzymes) or the bloodstream (hormones).

(b)   Motility; muscle cells are capable of contraction, using the energy of ATP to cause movement.

(c)   Photosynthesis; palisade cells are the site of much photosynthetic activity in the leaf (see Problem 9).

(d)   Absorption; intestinal mucosal cells are especially well-suited for this function because they contain microvilli that greatly increase the absorptive surface area of the cell (see Figure 4-2).

(e)   Transmission of impulses; the most important function of nerve cells is to conduct electrical signals from one part of the body to another.

(f)   Cell division; of the cell types listed, only bacterial cells are capable of rapid and repeated division, with only about 20-30 minutes between division under optimal conditions in at least some species.

4.   (a) P   (d) A, B, P   (g) A, P   (j) A, B, P

(b) B, P   (e) A, P   (h) P   (k) A, B, P

(c) A, P   (f) A, P   (i) P   (l) A, P

5.   (a)   Peptidoglycan; chief component.

(b)   Prokaryotes; blue-green algae are prokaryotes.

(c)   Hydrolases; lysosomal enzymes catalyze hydrolytic reactions.

(d)   Pores; the nuclear envelope is characterized by its pores.

(e) Ribosomes; the nucleolus is the site of assembly of ribosomal subunits.

(f) Secretory proteins; these proteins are synthesized by ribosomes bound to the rough ER.

(g) Lipid synthesis; occurs in smooth ER.

(h) Bacteriophage; viruses that infect bacterial cells.

6. (a) look for plastids or the central vacuole.

(b) become flaccid as water is drawn out of its cells.

(c) a ribosome, microtubule, microfilament, etc.

(d) they are not membrane-bounded.

(e) they are very similar in size.

7. (a) B   (d) N   (g) A

(b) N   (e) A   (h) N

(c) B   (f) N   (i) B

8. (a) Photorespiration; the chloroplast provides the glycolate which the peroxisome oxidizes.

(b) Processing of secretory proteins; processing is begun in the rough ER and continued in the Golgi complex.

(c) Protein synthesis; messenger RNA is synthesized in the nucleus and used to direct protein synthesis by ribosomes.

(d) Breakdown of sugar to obtain energy; begun in the cytoplasm, continued in the mitochondrion.

(e) Cellular secretion; secretory proteins are brought to the plasma membrane by secretory vesicles, which then fuse with the plasma membrane to expel their contents to the outside by exocytosis.

9. (a) Summary of calculations:

Structure	Dimensions	Volume	Number	Total Volume	Percent of Cell Volume
	μm	μm³		μm³	%
Cell	20×35	11,000	1	11,000	100.0
Vacuole	15×30	5300	1	5300	48.2
Nucleus	$d = 6$	113	1	113	1.0
Chloroplasts	2×8	25	40	1000	9.1
Mitochondria	1×2	1.570	200	314	2.9
Peroxisomes	$d = 0.5$	0.065	100	6.5	0.06
Ribosomes	$d = 0.025$	8.2×10^{-6}	2×10^6	16.3	0.14

(b) Volume remaining $= 100 - (48.2 + 1.0 + 9.1 + 2.9 + 0.06 + 0.14)$

$= 100 - 61.4 = 38.6\%.$

This volume must accommodate the Golgi complex, the endoplasmic reticulum, the cytoskeleton, etc.

10. (a) 3   (c) 5   (e) 4   (g) 2

(b) 7   (d) 1   (f) 6

11. Either viewpoint is defensible; the relevant arguments are discussed on p. 101.

5

Bioenergetics: The Flow of Energy in the Cell

1. (a) $(1.94 \text{ cal/min} \cdot \text{cm}^2)(5.26 \times 10^5 \text{ min/year})(1.28 \times 10^{18} \text{ cm}^2)$

    $= 1.3 \times 10^{24} \text{ cal/year.}$

    (b) Some incident radiation is reflected back into space and much is absorbed by components of the earth's atmosphere. Atmospheric ozone plays an important role in filtering out ultraviolet radiation, while water vapor is responsible for most of the absorption in the infrared range.

    (c) Much of the radiation falls on areas of the earth's surface where the climate is not favorable (too hot, too cold, too dry) for growth of phototrophic organisms during at least part of the year. In addition, about two-thirds of the earth's surface is covered by oceans which, though quantitatively significant in global photosynthesis, have in general only a very low density of phototrophic organisms and therefore a low efficiency of light utilization. Moreover, incident radiation represents a broad spectrum of wavelengths that can be used with varying degrees of efficiency by photosynthetic pigments, as discussed further in Chapter 12.

2. (a) Glucose is 40% carbon (72/180 = 0.4), so 5×10^{16} g carbon represents about 12.5×10^{16} g organic matter. (That is about 140 billion tons, if nonmetric units help you imagine the magnitude of the process.)

(b) $(12.5 \times 10^{16}$ g$)(3.8$ kcal/g$) = 4.75 \times 10^{17}$ kcal $= 4.75 \times 10^{20}$ **cal.**

(c) $(4.75 \times 10^{20}$ cal$)/(1.3 \times 10^{24}$ cal$) = 3.7 \times 10^{-4} = 0.037\%$. (This means that more total solar energy will be received by the earth during the next 12 months than has been trapped photosynthetically since 700 BC!)

(d)   Virtually all of it must be consumed by chemotrophs, since the earth is not undergoing any dramatic annual increase in amount of phototrophic organic matter accumulated.

3.   (a)   Solar energy $\longrightarrow$ absorbed by chlorophyll $\longrightarrow$ used to make ATP and high-energy reduced coenzyme $\longrightarrow$ sugar in leaf $\longrightarrow$ sugar in ear $\longrightarrow$ starch in ear $\longrightarrow$ starch in horse's gastro-intestinal tract $\longrightarrow$ sugar in horse's blood $\longrightarrow$ ATP in muscle cells $\longrightarrow$ muscle contraction.

(b)   The energy has been converted quantitatively to heat.

(c)   The corn plant also furnishes fixed carbon, organic nitrogen, and oxygen.

(d)   The horse furnishes carbon dioxide and nitrogenous wastes (not to mention horse manure!).

4.   (a)   Use of blood sugar as a source of energy for muscle contraction; important for animal motility.

(b)   Use of chemical energy to cause flash of light by firefly; important as mating signal.

(c)   Photosynthetic use of sunlight to synthesize sugar; entire biosphere depends on this process as its energy link with the sun.

(d)   Use of chemical energy to generate a potential and deliver an electric shock; defense mechanism of electric eel.

(e) Use of chemical energy to pump protons into stomach and maintain low pH; aid to digestion of foodstuffs.

5. Reactions tend to proceed toward thermodynamic equilibrium because the mixture of reactants and products present at equilibrium represents their lowest free-energy state and hence the most thermo-dynamically stable condition. A steady state exists when the reactants and products of a given reaction are maintained at concentrations far from their equilibrium state. The ability of a cell to maintain the reactants and products of many reactions far from their equilibrium concentrations ensures that the thermodynamic drive toward equilibrium can be used by the cell to perform needful work (i.e., effect needful changes), thereby maintaining and extending its activities and its structural complexity.

6. (a) $\Delta G^{\circ\prime} = -RT \ln K'_{eq} = (-1.987)(298) \ln (0.165) = -592(-1.802)$

$= +1067$ cal/mol.

To convert 1 mole of 3-phosphoglycerate to 2-phosphoglycerate under conditions such that the concentrations of both are main-tained constant at 1.0 M would require the input of 1067 cal of free energy; the reaction is endergonic under standard conditions.

(b) $\Delta G' = \Delta G^{\circ\prime} + (1.987)(298) \ln \dfrac{4.3 \times 10^{-6}}{61 \times 10^{-6}}$

$= +1067 + 592 \ln (0.0705) = 1067 + 592(-2.652)$

$= -500$ cal/mol.

To convert 1 mole of 3-phosphoglycerate to 2-phosphoglycerate under the prevailing concentrations in the red blood cell will result in the liberation of 500 cal of free energy; the reaction is exergonic under prevailing conditions.

(c) The concentration of 2-phosphoglycerate can rise only to the equilibrium value; if it goes above that, equilibrium will lie to the left, and the reaction will proceed in the reverse direction.

To calculate the equilibrium value:

$$K'_{eq} = 0.165 = \frac{[2\text{-phosphoglycerate}]}{61 \times 10^{-6}}$$

$$[2\text{-phosphoglycerate}] = (0.165)(61 \times 10^{-6}) = 10 \times 10^{-6} \text{ M}$$

$$= 10 \ \mu M.$$

(Alternatively, you can set $\Delta G'$ equal to zero and solve for [2-phosphoglycerate] using equation 5-17.)

7. (a) $\Delta G' = \Delta G^{\circ\prime} + 592 \ln \dfrac{[\text{glucose}][\text{phosphate}]}{[\text{glucose-6-phosphate}]}$

$$= -3300 + 592 \ln \frac{(5 \times 10^{-3})(5 \times 10^{-3})}{20 \times 10^{-6}}$$

$$= -3300 + 592 \ln (1.25) = -3300 + 592(0.223)$$

$$= -3300 + 132 = -3168 \text{ cal/mol} = -3.17 \text{ kcal/mole.}$$

(b) The reaction will proceed to the right under these conditions.

8. (a) Under standard conditions, the reaction is thermodynamically feasible in the direction written (i.e., to the right).

(b) $\Delta G^{\circ\prime} = -RT \ln K'_{eq} = -(1.987)(298) \ln (19)$

$$= -592(2.94) = -1740 \text{ cal/mol} = -1.74 \text{ kcal/mol.}$$

(c) With the reactant and product concentrations equal, the reaction is thermodynamically feasible to the right, because the free-energy change is the same as under standard conditions.

(d) $\Delta G' = \Delta G^{\circ\prime} + RT \ln \dfrac{[\text{glucose-6-phosphate}]}{[\text{glucose-1-phosphate}]}$

$$= -1.74 + RT \ln (1) = -1.74 + 0 = -1.74 \text{ kcal/mol.}$$

9. (a) Since $\Delta G^{\circ\prime} = 0$, the equilibrium constant must be 1.0. The reaction will therefore continue to the right until all species are present at equimolar concentrations of 0.005 M each .

(b) Assume x mol/liter of succinate react with x mol/liter of FAD to generate x mol/liter each of fumarate and $FADH_2$.

At equilibrium:

[succinate] = $0.01 - x$

[FAD] = $0.01 - x$

[$FADH_2$] = $0.01 + x$

[fumarate] = x

$$K'_{eq} = 1.0 = \frac{[\text{fumarate}][\text{FADH}_2]}{[\text{succinate}][\text{FAD}]} = \frac{(x)(0.01 + x)}{(0.01 - x)(0.01 - x)}$$

$$= \frac{0.01x + x^2}{0.0001 - 0.02x + x^2}$$

$0.01x + x^2 = 0.0001 - 0.02x + x^2$, so $0.03x = 0.0001$,

and therefore $x = 0.0033$.

Equilibrium concentrations are therefore:

[succinate] = 0.0067 M

[FAD] = 0.0067 M

[$FADH_2$] = 0.0133 M

[fumarate] = 0.0033 M.

(c) $\Delta G' = \Delta G^{\circ\prime} + 592 \ln \dfrac{[\text{fumarate}][\text{FADH}_2]}{[\text{succinate}][\text{FAD}]} = -1500$ cal/mol

$$= 0 + 592 \ln \frac{(2.5 \times 10^{-6})(5)}{[\text{succinate}]} = -1500 \text{ cal/mol}$$

$$\ln \frac{(12.5 \times 10^{-6})}{[\text{succinate}]} = -\frac{1500}{592},$$

so $\dfrac{(12.5 \times 10^{-6})}{[\text{succinate}]} = e^{-2.534} = 0.07934$

$12.5 \times 10^{-6} = 0.07934$ [succinate],

so [succinate] = $(12.5 \times 10^{-6})/0.07934 = 157 \times 10^{-6}$ M = **157 μM.**

10. (a) For the overall reaction A $\rightarrow$ D

$$K'_{AD} = \frac{[D]}{[A]} = \frac{[D][B][C]}{[A][B][C]} = \frac{[B][C][D]}{[A][B][C]} = K'_{AB} \cdot K'_{BC} \cdot K'_{CD}.$$

(b) $\Delta G^{\circ\prime}_{AD} = -RT \ln K'_{AD} = -RT \ln [K'_{AB} \cdot K'_{BC} \cdot K'_{CD}]$

$= -RT \ln K'_{AB} - RT \ln K'_{BC} - RT \ln K'_{CD} = \Delta G^{\circ\prime}_{AB} + \Delta G^{\circ\prime}_{BC} + \Delta G^{\circ\prime}_{CD}.$

(c) $\Delta G'_{AD} = \Delta G^{\circ\prime}_{AD} + RT \ln \dfrac{[D]}{[A]} = \Delta G^{\circ\prime}_{AD} + RT \dfrac{[D][B][C]}{[A][B][C]}$

$= \Delta G^{\circ\prime}_{AD} + RT \ln \dfrac{[B][C][D]}{[A][B][C]}$

$= \Delta G^{\circ\prime}_{AB} + \Delta G^{\circ\prime}_{BC} + \Delta G^{\circ\prime}_{CD} + RT \ln \dfrac{[B]}{[A]} + RT \ln \dfrac{[C]}{[B]} + RT \ln \dfrac{[D]}{[C]}$

$= \Delta G^{\circ\prime}_{AB} + RT \ln \dfrac{[B]}{[A]} + \Delta G^{\circ\prime}_{BC} + RT \ln \dfrac{[C]}{[B]} + \Delta G^{\circ\prime}_{CD} + RT \ln \dfrac{[D]}{[C]}$

$= \Delta G'_{AB} + \Delta G'_{BC} + \Delta G'_{CD}.$

11. No! The full expressions for E and ΔE in terms of G, P, V, S, and T are as follows:

$E = G - PV + TS$

$\Delta E = \Delta G - \Delta(PV) + \Delta(TS)$

The only conditions for which Equation 5-27 is valid are those for which pressure, volume, and temperature are constant, so that $\Delta(PV) = 0$ and $\Delta(TS)$ becomes TΔS.

6

Enzymes:
The Catalysts of Life

1. (a) It means that the molecules that ought to react would release energy if they were to do so, but do not possess enough energy to collide in a way that allows the reaction to be initiated.

   (b) Touching a match to a sheet of paper is an example. Thermal activation imparts sufficient kinetic energy to the molecules such that the proportion of them possessing adequate energy to collide and react increases significantly. Once initiated, the reaction is self-sustaining, since reacting molecules release sufficient energy to energize and activate neighboring molecules for reaction.

   (c) Interaction of molecules is facilitated (by positioning on the catalyst surface, for example), thereby requiring less energy to activate each molecule and so ensuring that substantially greater numbers of molecules possess adequate energy to initiate reaction without any elevation in temperature.

   (d) Advantages: specificity, more exacting control.
   Disadvantages: much more susceptible to inactivation by heat, extremes of pH, and so on; also, much energy needs to be expended to synthesize the enzyme molecules.

2. (a) P   (c) C   (e) P   (g) C

   (b) N   (d) C   (f) P   (h) N

3.   (a)   In the absence of a catalyst, molecules of A can acquire enough
          kinetic energy to be converted to B, but not to C. Given enough
          time, therefore, compound B will predominate. Neither compound
          A nor compound C will be present in significant quantities,
          because B cannot be converted to either A or C and A cannot be
          converted to C, given the activation energy barriers for each of
          these potential reactions.

     (b)   In the presence of an appropriate catalyst, compound C will
          eventually predominate, because the activation energy barrier for
          the A → C conversion in the presence of the catalyst is lower than
          the activation energy barrier for the uncatalyzed A → B
          conversion.

4.   (a)   The temperature curve agrees well with the body temperature of
          most warm-blooded organisms, as well as with the approximate
          maximum environmental temperature for many other organisms.
          The pH plot agrees well with the approximately neutral pH of the
          cytoplasm and of some (though not all) of the subcellular
          compartments of most cells.

     (b)   Up to about 35°C, the temperature curve reflects the thermal
          activation expected of a chemical reaction. Above about 40°C,
          activity decreases rapidly as the structure of the protein is
          disrupted by denaturation. In the case of the pH plot, the decrease
          noted on either side of its optimum presumably reflects titration of
          charged groups on either the enzyme, the substrate, or both.
          Specifically, activity loss as the solution is made more acidic is due
          to protonation of chemical groups that must be ionized for
          maximum activity, whereas activity loss on the basic side of the
          curve is due to ionization of chemical groups that must be
          protonated for maximum activity.

     (c)   A plot of enzyme activity versus temperature is likely to peak at a
          much higher temperature for an enzyme from such a thermophilic

bacterium. The tertiary structure of such enzymes must be very stable to heat.

(d) *Pepsin* is a protein-degrading enzyme found in the stomach. It has a pH optimum at about 2, consistent with the acidic environment of the stomach. Pepsin presumably has chemical groups at its active site that must be protonated if the enzyme is to be active. As the pH increases, these groups are titrated to the ionized form, resulting in inactivation of the enzyme. *Cholinesterase* is an enzyme involved in transmission of nerve impulses at the junctions between nerve cells. Its pH curve is consistent with the neutral or slightly alkaline pH of the nerve junction. Cholinesterase presumably has chemical groups at its active site that must be ionized for enzyme activity, with full ionization achieved at about pH 8. *Papain* is a protein-degrading enzyme present in the papaya plant. The flat pH curve means that the organism can make use of the enzyme over a broad pH range. The lack of pH effect on enzyme activity suggests that the active site does not contain charged amino acids whose ionization states are crucial to enzyme activity.

5. (a) A proteolytic enzyme is specific for a particular kind of bond *between* amino acids and is apparently little influenced by the exact chemical nature of the amino acid side groups surrounding that bond. A dehydrogenase attacks a specific bond *within* a molecule and apparently has an active site that is very specific for the particular molecule rather than the particular bond.

(b) The subtilisin active site probably recognizes only elements of the peptide bond itself, whereas that of trypsin must have a binding site that is specific for a positively charged amino acid.

(c) The structural analogues can bind to the active site, thereby rendering it unavailable for true substrate and as a result inoperative in catalysis. In effect, they render inoperative every active site they occupy for as long as they remain there.

6. (a) Induced fit refers to the conformational change in an enzyme that occurs when the appropriate substrate binds to the active site of the enzyme. This conformational change greatly enhances the specificity of the enzyme-substrate binding and positions the proper reactive groups of the active site maximally for catalysis. In the carboxypeptidase mechanism, induced fit occurs in step 1 when the negatively charged carboxylate group of the substrate interacts with the positively charged side chain of an arginine at the active site, thereby causing the active site to "close" around the substrate.

    (b) Electrophilic substitution occurs when an electropositive prosthetic group at the active site of an enzyme accepts one or more electrons from an electronegative group on the substrate. The carboxy-peptidase reaction mechanism does not involve electrophilic substitution.

    (c) Nucleophilic substitution occurs when an electronegative group at the active site of an enzyme donates one or more electrons to an electropositive group on the substrate. This occurs at step 4 in the carboxypeptidase mechanism, when the carbon atom of the polarized carbonyl group of the substrate accepts an electron from the electronegative carboxylate group of the glutamate at the active site.

    (d) Proton donation occurs when a proton is transferred from an ionizable group at the active site of an enzyme to the substrate molecule. In the carboxypeptidase mechanism, this occurs at step 4 with the transfer of a proton from the hydroxyl group of a tyrosine to the amide nitrogen of the substrate.

7. (a) C   (c) B   (e) C

    (b) A   (d) B   (f) A

8.   (a)   If product accumulation becomes significant, the back-reaction may begin to occur, and the net activity in the forward direction will be correspondingly reduced.

(b)   See the graph in Figure A6-1 (solid line). Doubling of substrate concentration always results in less than a twofold increase in velocity because the curve follows a hyperbolic equation, such that the relationship between substrate concentration and velocity is not linear at any point.

(c)   Results with 0.5 µg enzyme per tube are shown as the dashed line on the graph of Figure A6-1. Half as much enzyme results in half the total rate of lactose hydrolysis at any concentration of lactose (that is, reaction velocity is linear with enzyme concentration).

(d)   The data for the double-reciprocal plot are calculated as shown in Table A6-1. For the double-reciprocal plot, see the graph in Figure A6-2 (solid line).

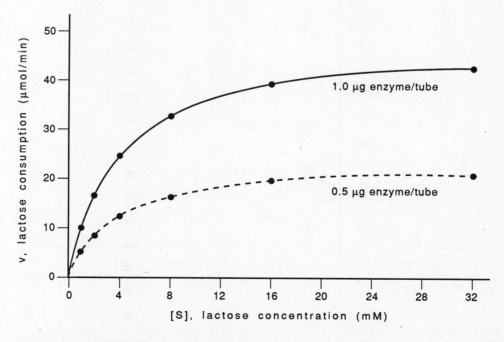

Figure A6-1.   Kinetics of the β-Galactosidase Reaction. The dependence of initial reaction velocity on substrate concentration is shown for the data of Problem 8 with two concentrations of enzyme: 1.0 µg per tube (solid line), and 0.5 µg per tube (dashed line).

[S] mM	1/[S] 1/mM	v μmol/min	1/v min/μmol
1.0	1.000	10.0	0.100
2.0	0.500	16.7	0.060
4.0	0.250	25.0	0.040
8.0	0.125	33.3	0.030
16.0	0.0625	40.0	0.025
32.0	0.03125	44.4	0.022

Table A6-1. Double-Reciprocal Data for Problem 8.

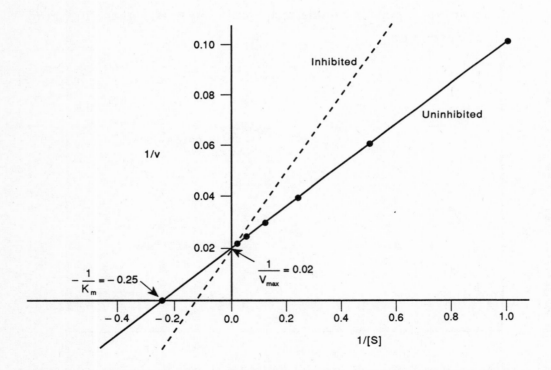

Figure A6-2. Double-Reciprocal Plot for the β-Galactosidase Reaction. Reciprocal values for the initial reaction velocity v and the substrate concentration [S] were determined for the data of Problem 8 and plotted as 1/v vs. 1/[S] (solid line). In the presence of a competitive inhibitor that increases the apparent K_m value by a factor of two, the line will be twice as steep (dashed line).

(e) K_m: X-intercept is –0.25, so K_m = –1/–0.25 = + 4.0 mM.

   V_{max}: Y-intercept is 0.02, so V_{max} = 1/0.02 = 50 μmoles/min.

(f)   The expected result is shown as a dashed line on the graph in Figure A6-2.

9.   (a)   Figure A6-3a shows the Michaelis-Menten plots for both glucokinase and hexokinase. The double-reciprocal plots are shown in Figure A6-3b.

   (b)

Blood Glucose Level		Glucokinase Activity		Hexokinase Activity	
mg/100 ml	mM	μmol/min	% of total	μmol/min	% of total
(i) 80	4.44	0.461	83%	0.096	17%
(ii) 120	6.67	0.600	86%	0.098	14%

(c)   Glucokinase accounts for more than 80% of the total glucose phosphorylation activity in the liver under both conditions, mainly because its V_{max} value is 15 times greater than that of hexokinase.

(d)   Glucokinase responds the most to increases in blood glucose level because the level of glucose in the blood is in the same range as the K_m of the enzyme; thus, the Michaelis-Menten curve is still rising rapidly in this region. Increasing blood glucose 50% (from 80 to 120 mg/100 ml) results in a 30% increase in glucokinase activity. Hexokinase, however, is essentially saturated in this range and increases less than 2% as blood glucose rises from 80 to 120 mg/100 ml.

(e)   Since glucokinase is quantitatively the most significant enzyme for glucose phosphorylation and also possesses the ability to respond to increases in blood glucose level, the ability of the liver to metabolize glucose is seriously impaired in the absence of glucokinase.

(f)   Since it plays only a minor role in glucose phosphorylation but is also active with a variety of hexoses, hexokinase may be more important in the phosphorylation of other hexoses, especially those present in lower concentrations.

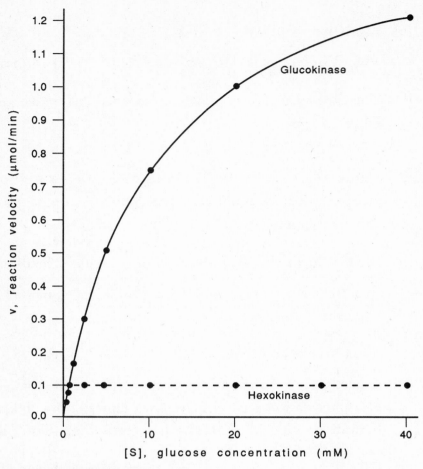

(a) Michaelis-Menten plot

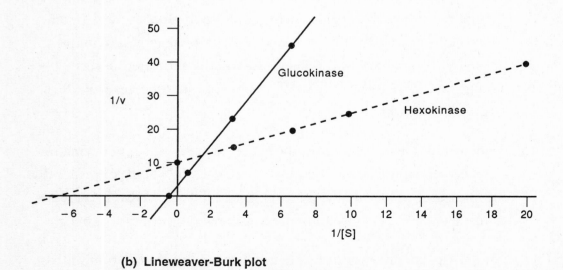

(b) Lineweaver-Burk plot

Figure A6-3.   Kinetics of the Glucokinase and Hexokinase Reactions.
(a) Michaelis-Menten plots for the glucokinase and hexokinase data of Problem 9.
(b) Lineweaver-Burk (double-reciprocal) plot for the same data.

10. (a) Mutant 1: K_m is higher, V_{max} is higher.

    Mutant 2: K_m is the same, V_{max} is lower.

(b) Since mutant 2 has a lower V_{max} value than the normal enzyme, its reaction velocity at any given substrate concentration will be lower than the velocity for the normal enzyme at the same substrate concentration.

(c) Mutant 1 is somewhat less efficient than the normal enzyme, because its K_m value is higher relative to the normal enzyme than is its V_{max} value, resulting in a lower V_{max}/K_m ratio. This can be seen readily by noting that the slope of the line for mutant 1 is greater than that for the normal enzyme. Recalling (from Figure 6-13) that the slope is K_m/V_{max}, it follows that a mutant with a higher slope will have a lower efficiency, since efficiency is defined as the inverse of the slope.

11. (a)

Eadie-Hofstee Plot	**Hanes-Woolf Plot**
Cross-multiplication of the Michaelis-Menten expression yields	Cross-multiplication of the Michaelis-Menten expression yields
$vK_m + v[S] = V_{max}[S]$.	$V_{max}[S] = vK_m + v[S]$.
Dividing by [S] gives	Dividing by $v \cdot V_{max}$ gives
$vK_m/[S] + v = V_{max}$, so	$[S]/v = K_m/V_{max} + (1/V_{max})[S]$.
$v = V_{max} - K_m \, (v/[S])$.	

(b)

Plot v versus $v/[S]$.	Plot $[S]/v$ versus $[S]$.
Y-intercept = V_{max}	X-intercept = $-K_m$
Slope = $-K_m$	Slope = $1/V_{max}$
X-intercept = V_{max}/K_m	Y-intercept = K_m/V_{max}

(c)   The Hanes-Woolf plot has the data points spaced out linearly along the X-axis, as in the original Michaelis-Menten plot. Both of the other plots use the reciprocal of [S] for the X-axis, thereby giving too much weight to values determined at low substrate concentrations.

7

Membranes: Their Structure and Chemistry

1. (a) Detection and recognition.   (f) Anchoring of function.

    (b) Control of movement.   (g) Compartmentalization.

    (c) Anchoring of function.   (h) Control of movement.

    (d) Detection and recognition.   (i) Detection and recognition.

    (e) Control of movement.

2. (a) Evidence of membrane asymmetry; 1950s.

    (b) Evidence that the membrane permeability of solutes is related to how nonpolar they are; 1890s.

    (c) Evidence that the phospholipid bilayer is not completely covered by surface layers of protein; 1960s.

    (d) Evidence that such particles, when seen on biological membranes, are integral proteins; 1970s.

    (e) Evidence that real membranes are not just phospholipid bilayers; 1920s.

    (f) Evidence that the thin, electron-dense lines of real membranes cannot be explained only in terms of protein staining; 1960s.

    (g) Evidence of two categories of membrane proteins differing significantly in location within the membrane and/or in their affinity for an aqueous environment; 1970s.

(h) Evidence that light-dependent proton pumping can be carried out by a relatively simple protein-pigment complex; in addition, the crystalline array allowed electron diffraction analysis, which led to the first understanding of how a membrane protein is organized within the lipid bilayer; 1970s.

3. (a) To calculate the number of protein molecules in the membrane, determine the number of moles and multiply by Avagadro's number: $(6.0 \times 10^{-13} \text{ g})(1 \text{ mole}/50,000 \text{ g})(6.023 \times 10^{23})$ $= 7.2 \times 10^6$ **protein molecules/membrane.**

(b) Let p = number of grams of phospholipid in one membrane and c = number of grams of cholesterol in one membrane. Then $p + c = 5.2 \times 10^{-13}$ g.

Because phospholipid and cholesterol are present in a 1:1 molar ratio, the number of moles of phospholipid (p/800) must equal the number of moles of cholesterol (c/386): p/800 = c/386. Solving these two simultaneous equations, we get:

$p = 3.5 \times 10^{-13}$ g of phospholipid per membrane

$c = 1.7 \times 10^{-13}$ g of cholesterol per membrane.

From these values we can calculate that a red cell membrane contains 2.6×10^8 molecules each of phospholipid and cholesterol, for a total of 5.2×10^8 lipid molecules per membrane. The lipid/protein ratio is therefore $(5.2 \times 10^8)/(7.2 \times 10^6) =$ **72 lipid molecules/protein molecule.** (This is a very representative value for the lipid/protein ratio of most plasma membranes.)

(c) The phospholipid molecules occupy a total area of $(2.6 \times 10^8$ molecules)$(0.55 \text{ nm}^2/\text{molecule}) = 1.54 \times 10^8 \text{ nm}^2$. Similarly, the cholesterol molecules occupy a total area of $(2.6 \times 10^8$ molecules)$(0.38 \text{ nm}^2/\text{molecule}) = 1.06 \times 10^8 \text{ nm}^2$. Phospholipid and cholesterol therefore account for a surface area of 2.6×10^8 nm^2. The membrane is a bilayer, so this surface area is divided between two monolayers. The surface area per monolayer is therefore 1.3×10^8 nm^2. Since 1 nm^2 = 1×10^{-6} μm^2, the surface area occupied by lipid molecules can be expressed as 130 μm^2. This represents $130/145 \times 100\%$ = 90% **of the total surface area of the membrane.** (The remaining 10% of the surface area is occupied by transmembrane proteins.)

4. (a) Lateral diffusion occurs readily because it does not require polar portions of molecules to move through a nonpolar environment or vice versa; molecules simply move in the semifluid phospholipid bilayer while maintaining their most stable thermodynamic shape and orientation. Transverse diffusion requires that polar portions of molecules pass through the hydrophobic interior of the membrane, a thermodynamically unfavorable event.

    (b) Membrane components are free to distribute themselves throughout the phospholipid layer into which they were originally inserted; only those components that are externally anchored can be expected to maintain a fixed position in the membrane.

    (c) The asymmetry that is so characteristic and functionally important a feature of membranes would not be possible, because components would readily and rapidly equilibrate between the inner and outer phospholipid layers.

5. (a) Yes; fluidity is temperature-dependent.

    (b) No; longer-chain fatty acids decrease the fluidity.

    (c) Yes; unsaturated fatty acids increase fluidity.

    (d) Yes; unsaturated fatty acids increase fluidity.

    (e) No; no effect of temperature change on protein content.

6. (a) D, T, G   (c) T   (e) None   (g) D

    (b) G, P   (d) G   (f) T

7. (a) Myelin: $\dfrac{(0.25)(1.33) + (1.0)(0.92)}{0.25 + 1.0} = 1.00 \text{ g/cm}^3$.

    Inner membrane: $\dfrac{(3.6)(1.33) + (1.0)(0.92)}{3.6 + 1.0} = 1.24 \text{ g/cm}^3$.

    Chloroplast: $\dfrac{(0.8)(1.33) + (1.0)(0.92)}{0.8 + 1.0} = 1.10 \text{ g/cm}^3$.

    Plasma membrane: $\dfrac{(1.2)(1.33) + (1.0)(0.92)}{1.2 + 1.0} = 1.14 \text{ g/cm}^3$.

    (b) $\dfrac{(1.33)(R) + (1.0)(0.92)}{R + 1.0} = 1.18 \text{ g/cm}^3$; R = 1.73.

    (c) Each organelle has a characteristic density, based in part on the protein/lipid ratio of its membrane(s), and will migrate to that position when centrifuged to equilibrium on a density gradient. Mitochondria band at a higher density than do chloroplasts because of the high protein content of their inner membrane. To band at a density of 1.25 g/cm³, lysosomes presumably have a protein/lipid ratio about that of the mitochondrial inner membrane (3.6).

8. (a) Some of the membrane proteins are associated with the outer phospholipid layer of the plasma membrane and protrude out from the membrane sufficiently to allow exposed tyrosine groups to be labeled by the LP reaction.

    (b) Some of the membrane proteins associated with the outer phospholipid layer of the plasma membrane are glycoproteins, the carbohydrate side chains of which are accessible to the GO and borohydride.

(c) All the glycoproteins of the erythrocyte membranes are associated with the outer phospholipid layer of the membrane, and at least a portion of every carbohydrate side chain protrudes from the membrane surface sufficiently far to be labeled.

(d) All the proteins associated with the outer phospholipid layer of the plasma membrane are glycoproteins; proteins bearing no carbohydrate side chains are, without exception, inaccessible to the labeling reagents.

(e) All major membrane proteins protrude at least to some extent on one side of the membrane or the other; none is totally buried in the interior of the membrane.

9. (a) You would expect no labeling, since we already know from Problem 8(c) that all glycoproteins are associated with the outer layer and would therefore be on the interior of an inside-out vesicle.

(b) You would expect to see labeling of all the proteins that were not labeled in Problem 8(a), since we know from Problem 8(e) that almost all membrane proteins are accessible from one side of the membrane or the other.

(c) You would conclude that at least some proteins extend all the way through the membrane and actually protrude sufficiently on both sides of the membrane to allow them to be labeled on either side.

10. Label membrane proteins of intact cells with one of the techniques described; then prepare inside-out vesicles and use the second labeling technique.

8

Transport Across Membranes: Overcoming the Permeability Barrier

1. (a) P, A   (c) D   (e) A   (g) N   (i) D, P, A,

   (b) D   (d) N   (f) D   (h) D, P   (j) P, A

2. (a) The relationship between v and C is linear for ethanol, but hyperbolic for acetate. Ethanol therefore crosses the membrane by diffusion, whereas acetate requires a carrier.

   (b) Ethanol is not charged and is known to be miscible with organic solvents, whereas acetate is negatively charged at neutral pH and is water-soluble.

   (c) From the graph of v versus C for acetate uptake, the V_{max} value appears to be about 100 µmol/min, so the K_m value must be about 1 mM, since that is the concentration at which the velocity is 50 µmol/min.

   (d) Yes for ethanol, since the data exclude a carrier-mediated process; no for acetate, since we do not know the internal concentration of acetate.

3. (a) The concentration gradient is $10^{-2}/10^{-7} = 10^5$.

(b) $\Delta G_{outward} = RT \ln \dfrac{[H^+]_{out}}{[H^+]_{in}} - zEF$

$= (1.987)(310) \ln (10^5) - (+1)(-0.07)(23,062)$

$= + 7092 \text{ cal/mol} + 1613 \text{ cal/mol}$

$= + 8705 \text{ cal/mol} = \textbf{+8.7 kcal/mol.}$

(c) $\Delta G°$ for ATP hydrolysis is only -7.3 kcal/mol, but concentrations of ATP, ADP, and P_i are usually such that ΔG for ATP hydrolysis in the cell is in the range of -10 to -14 kcal/mol, which would probably be just barely adequate to drive hydrogen secretion.

(d) If the membrane potential (E) is to be just high enough to prevent the inward movement of protons when the outside-to-inside proton gradient is 10^5, then we can write

$\Delta G_{inward} = RT \ln \dfrac{[H^+]_{in}}{[H^+]_{out}} + zEF = 0$

or $E = (-RT/zF) \ln (10^{-5})$

$= -(1.987)(310)/(+1)(23,062) \times \ln (10^{-5})$

$= (-0.0267)(-11.5) = \textbf{+0.308 V = +308 mV.}$

4. (a) Such vesicles are free of endogenous energy sources and do not metabolize most substrates. However, you cannot be sure that a given transport system will be active in such vesicles, and it is possible that membrane proteins could alter their positions during vesicle formation.

(b) Na^+:outside; K^+:inside; ATP:outside

(c) ATP hydrolysis will continue at a high constant rate until the sodium and potassium concentration gradients across the membrane approach the maximum levels attainable with the given concentrations of ATP, ADP, and P_i. The rate will then drop rapidly to the low baseline level of whatever ATP hydrolysis is necessary to replenish the gradient due to ion leakage across the membrane.

5. (a) Figure 8-14 shows that 0.2 µmoles of ATP were hydrolyzed during an interval of 1 minute (beginning at 2 minutes and ending at 3 minutes) by 1.0 mg of protein present as added sarcoplasmic reticulum. The ATPase activity is therefore **0.2 µmoles/min per milligram.**

   (b) During the same 1-minute interval, all of the added calcium (0.4 µmoles) was taken up, as shown by the change in slope of the ATP hydrolysis rate at the end of the minute. The Ca^{2+}/ATP ratio is therefore 0.4 µmoles/0.2 µmoles = **2 calcium ions transported inward for each molecule of ATP that is hydrolyzed.**

   (c) When the ionophore is added at 4 minutes, the calcium that had been accumulated within the vesicles leaks out. The presence of calcium ions in the medium activates the calcium-dependent ATPase, so that ATP hydrolysis begins once again. The reaction will continue until all of the ATP has been hydrolyzed to ADP and phosphate.

6. (a) ΔG_{inward} = RT ln $\dfrac{[Na^+]_{in}}{[Na^+]_{out}}$ + zEF

   = (1.987)(310) ln (7.5/105) + (+1)(−0.065)(23,062)

   = 616 ln (0.0714) − 1498 = −1626 − 1498

   = −3124 cal/mol = **−3.12 kcal/mol.**

   (b) ΔG_{inward} = RT ln $\dfrac{[glycine]_{in}}{[glycine]_{out}}$ + zEF

   = (1.987)(310) ln (15/0.1) + 0 (no net charge)

   = 616 ln (150) = +3086 cal/mol

   = **+3.09 kcal/mol.**

Inward glycine transport requires the concomitant inward transport of at least one sodium ion.

(c) $\Delta G_{inward} = RT \ln \dfrac{[\text{aspartate}]_{in}}{[\text{aspartate}]_{out}} + zEF$

$= (1.987)(310) \ln (22.5/0.15) + (-1)(-0.065)(23,062)$

$= 616 \ln (150) + 1498 = +3086 + 1498$

$= +4584 \text{ cal/mol} = \mathbf{+4.58 \text{ kcal/mol}}.$

Inward aspartate transport requires the concomitant inward transport of at least 2 sodium ions.

(d) The inward transport of aspartate requires more energy because, unlike glycine, aspartate is negatively charged at pH 7, so that both the aspartate concentration gradient and the membrane potential oppose uptake of this amino acid.

(e) Neither aspartate nor glycine is likely to diffuse freely back through the epithelial cell membrane, because both are highly polar molecules.

7. (a) For ATP hydrolysis:

$\Delta G = -7300 + RT \ln \dfrac{[\text{ADP}][P_i]}{[\text{ATP}]}$

$= -7300 + (1.987)(298) \ln (0.002)(0.001)/(0.020)$

$= -7300 + 592 \ln (0.0001) = -7300 + (592)(-9.210)$

$= -7300 + (-5452) = -12,752 \text{ cal/mol ATP}.$

Available per sodium ion:

$\Delta G = (-12,752 \text{ cal/mol ATP})(1 \text{ mol ATP}/3 \text{ mol Na}^+)$

$= -4,250 \text{ cal/mol Na}^+$

For outward sodium movement:

$$\Delta G_{outward} = RT \ln \frac{[Na^+]_{out}}{[Na^+]_{in}} - zEF$$

$$= 592 \ln \frac{0.15}{[Na^+]_{in}} - (+1)(-0.075)(23,062)$$

$$= 592 \ln \frac{0.15}{[Na^+]_{in}} + 1730 = +4250 \text{ cal/mol.}$$

$$\ln \frac{0.15}{[Na^+]_{in}} = (4250 - 1730)/592 = 2520/592 = 4.26$$

$$\frac{0.15}{[Na^+]_{in}} = e4.26 = 70.8$$

$[Na^+]_{in} = 0.15/70.8 = 0.00212 = \textbf{2.12 mM.}$

(b) If it were an uncharged molecule, the internal concentration could be reduced much more because the ATP-driven pumping would not have to "fight" the membrane potential.

8. (a) No; sodium ion cotransport is not required for passive transport.

(b) Yes; cotransport of sodium ions drives the inward movement of amino acids and can only occur if sodium ions are actively pumped back out again.

(c) Yes; potassium ions must be actively pumped into red blood cells, and this can only occur via a pump that couples the inward pumping of potassium ions to the outward pumping of sodium ions.

(d) No; active uptake of sugars and amino acids in bacteria does not appear to be coupled to sodium cotransport as it is in animal cells.

9. (a) Roots were probably excised from young barley seedlings and suspended in a solution of potassium chloride containing radioactive potassium at the desired concentration. At the end of a uniform exposure period, the roots were rinsed in nonradioactive potassium chloride solution to remove $^{42}K^+$ not actually inside the cells and then assayed for radioactivity. The data were expressed as counts per minute taken up per gram of root tissue, converted to micromoles (from the known specific activity of the potassium chloride solution) and corrected for exposure time, if greater than 1 hour.

(b) Component 1: Y-intercept is about 0.07, so $V_{max(1)}$ is about 14.3 μmol/g-h; X-intercept is about –50, so $K_{m(1)}$ is about 0.02 mM.

Component 2: Y-intercept is about 0.08, so $V_{max(2)}$ is about 12.5 μmol/g-h; X-intercept is about –0.10, so $K_{m(2)}$ is about 10.0 mM.

(c) $$v_T = v_1 + v_2 = \frac{V_{max(1)}[S]}{K_{m(1)} + [S]} + \frac{V_{max(2)}[S]}{K_{m(2)} + [S]}$$

$$= \frac{14.3[S]}{0.02 + [S]} + \frac{12.5[S]}{10.0 + [S]}$$

Choose a series of values for S, plug into the above equation, calculate a series of corresponding v_T values, and plot v_T versus S on a scale with a discontinuity between 0.2 and 10 mM on the X-axis. There should be close agreement with the biphasic curve of Figure 8-15.

(d) Sodium appears to inhibit the transport mechanism responsible for component 1 of the uptake curve only minimally, and the inhibition is competitive. Sodium is a very potent inhibitor of the mechanism responsible for component 2, and the inhibition is probably competitive in this case also.

9
Intracellular Compartments

1. (a) Smooth ER; drug detoxification.

    (b) Rough ER and Golgi; secretion.

    (c) Smooth ER; steroid synthesis.

    (d) Rough ER and Golgi; protein processing.

    (e) Either peroxisome or smooth ER; detoxification.

    (f) Lysosome; autophagy.

2. (a) RS   (c) S   (e) R   (g) RS

    (b) R   (d) S   (f) S   (h) R

3. Incubate cultured bone marrow cells briefly with ^{35}S-labeled inorganic sulfate ($^{35}SO_4^{2-}$), followed by exposure to a large excess of unlabeled sulfate. At time intervals thereafter, homogenize aliquots of the cells and subject the homogenates to differential centrifugation to resolve rough ER, Golgi, secretory granules, and other subcellular structures. Assay each fraction for radioactivity and plot the amount of ^{35}S in each compartment as a function of time. If the sulfotransferase activity is in fact located in the Golgi, label will appear in that fraction first, and its disappearance from the Golgi will be accompanied by the progressive appearance of label in the secretory granules. Eventually, this should be followed by the appearance of ^{35}S-labeled chondroitin sulfate in the

medium. The experiment could also be done autoradiographically, in which case exposed silver grains would first appear over the Golgi, then over the secretory granules.

4. For the synthesis and glycosylation of glycoproteins of the plasma membrane, see Figure A9-1. Glycoproteins are found only in the outer

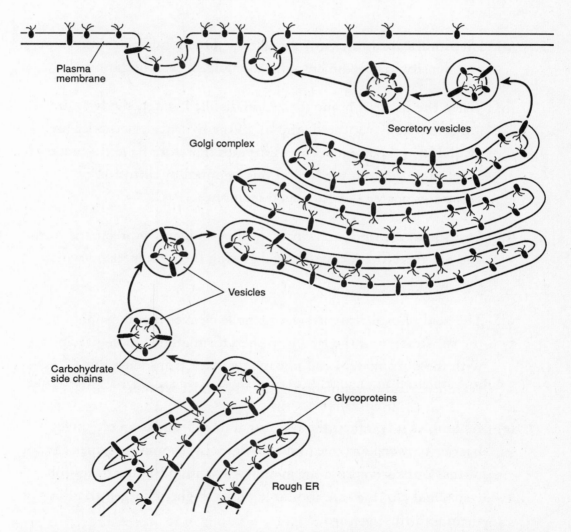

Figure A9-1. Synthesis and Glycosylation of Integral Membrane Proteins of the Plasma Membrane. Integral membrane proteins are synthesized on the rough ER, and oligosaccharide side chains are added in part on the luminal side of the rough ER (core glycosylation) and in part on the luminal side of the Golgi complex (terminal glycosylation). Side chains therefore face the interior of both organelles as well as the interior of transport vesicles, and become oriented toward the exterior of the cell when the appropriate vesicles fuse with the plasma membrane.

phospholipid layer of the plasma membrane because this is the layer that originally faced the interior of the rough ER and Golgi vesicles, where the enzymes involved in glycosylation are located. This assumes (a) that membrane asymmetry is maintained throughout the rough ER, Golgi, and plasma membrane, and (b) that the relationship of these membrane systems to each other is essentially as illustrated in the figure.

5. (a)   Lysosomes are sufficiently similar to mitochondria in size, shape, and density to cosediment with them upon centrifugation.

    (b)   When the tissue is homogenized in distilled water, the lysosomes are osmotically ruptured, and all of the hydrolase activities are available in the homogenate. Enzymes therefore have free access to substrates, without the limitations imposed by diffusion of substrates across the lysosomal membrane.

    (c)   Lysosomal enzymes are synthesized in an inactive precursor form and are therefore incapable of digesting the structures on which they are made.

    (d)   The acid phosphatase in the organelle cleaves the phosphate from the substrate, and the lead ions then form an insoluble precipitate with the phosphate. Lead phosphate is electron-dense and therefore readily visible in the electron microscope.

    (e)   As long as it is intact, the lysosomal membrane is an effective barrier between enzyme and substrate. Lysosomal enzymes cannot diffuse outward across the membrane, and most substrates for lysosomal enzymes are too large or too highly charged to allow them to diffuse inward.

6. (a) A, B   (f) A, B, D

    (b) A, B, C, D   (g) B, C

    (c) C   (h) D

    (d) D   (i) A, B

    (e) A, B   (j) C

7. (a) The fibers or particles are probably taken up by endocytosis,
         followed by fusion between the resulting heterophagic vacuoles
         and primary lysosomes to form heterophagic lysosomes.

    (b) The fibers or particles may physically abrade the lysosomal
         membrane, causing it to become leaky.

    (c) Death of the cells is probably due to escape of hydrolases from
         damaged secondary lysosomes.

    (d) The fibers or particles released upon cell death presumably are
         available to be ingested by other macrophages, with the same end
         result. Since the fibers are not digestible and no means exists to
         remove them from the lungs, a cycle of uptake, lysosomal damage,
         cell death, and fiber release is set up that can continue indefinitely,
         killing more and more cells.

    (e) Apparently, exposure to silica particles causes the macrophages to
         produce and release some sort of soluble factor capable of
         stimulating fibroblast cells in the lung to deposit collagen fibers,
         probably as an attempt to "wall off" and thereby to contain the
         silica in the lungs.

8. (a) S   (c) S   (e) C   (g) N

    (b) C   (d) N   (f) S   (h) C

9. (a) Minimal criteria would be a set of enzymes different from that of any other type of organelle and a means of separating the new set of enzymes (and hence the new organelle) from all other organelles or cell structures. The most generally useful means of achieving such separation is differential centrifugation followed by equilibrium density centrifugation.

    (b) Yes, provided that equilibrium density centrifugation was part of the procedure used to achieve separation.

    (c) Probably not, since peroxisomes and lysosomes are usually very similar in size, shape, and density.

10

Energy from Chemical Bonds: The Anaerobic Mode

1. (a) The clue is in the word *zymase*, an early term for enzyme. The heat-labile fraction (zymase) contains the enzymes and the heat-stable fraction (cozymase) contains the coenzyme (NAD^+) necessary for fermentative activity. This observation is important in distinguishing enzymes from coenzymes and understanding the need for both.

   (b) Phosphate is required as substrate in Reaction Gly-6, and the sequence cannot function without it. This observation is important in establishing the involvement of phosphate groups in energy metabolism.

   (c) Iodoacetate is an inhibitor of aldolase, the enzyme that splits fructose-1,6-bisphosphate into two trioses; in the presence of iodoacetate, fructose-1,6-bisphosphate accumulates. This observation is important to establish fructose-1,6-bisphosphate as an intermediate in the pathway.

   (d) Fluoride ion is a potent inhibitor of enolase, the enzyme that catalyzes the removal of water from 2-phosphoglycerate to generate phosphoenolypyruvate in Reaction Gly-9. Addition of fluoride to the yeast extract caused the accumulation of two phosphorylated three-carbon acids, which could then be identified as 2- and 3-phosphoglycerate, thereby establishing the chemical nature of two further intermediates in the pathway.

2.   (a)   T; the carbon dioxide arises from carbon atoms 3 and 4 of glucose.

       (b)   F; it will be in the methyl carbon.

       (c)   T; carbon atoms 1 and 6 become indistinguishable at step Gly-5.

       (d)   F; the carbon atom bearing the phosphate group will be labeled.

3.   (a)   The glycosidic bond that links successive glucose units together in a polysaccharide has sufficient free energy of hydrolysis to allow it to be cleaved by phosphorolysis with the direct uptake of inorganic phosphate. The product is therefore a glucose molecule that is already phosphorylated (on carbon atom 1), which means that step Gly-1 of the glycolytic pathway is bypassed and the ATP that would otherwise be required there is saved.

       (b)   Since monosaccharide units are cleaved from a polysaccharide by phosphorolysis, we can suggest the same mechanism for sucrose. Phosphorolysis of a disaccharide will yield one phosphorylated monosaccharide that will yield three molecules of ATP by the glycolytic pathway (because step Gly-1 is bypassed and one less ATP molecule is needed) and one free monosaccharide that will require step Gly-1 and will therefore yield two molecules of ATP. In fact, sucrose phosphorylase yields glucose-1-phosphate and free fructose.

       (c)   Raffinose has three monosaccharides linked together by two glycosidic bonds. Phosphorolysis of these bonds generates two phosphorylated hexoses (yield: 3 ATP each) and one free hexose (yield: 2 ATP). The average ATP yield per monosaccharide is therefore (3 + 3 + 2)/3 = **2.67**.

4. (a) $\Delta G' = \Delta G^{\circ\prime} + RT \ln \dfrac{[\text{glucose - 6 - phosphate}]}{[\text{glucose}]\,[P_i]}$

$= +\,3300 + (1.987)(298) \ln \dfrac{0.08 \times 10^{-3}}{[\text{glucose}](1.0 \times 10^{-3})}$

$= +\,3300 + 592 \ln \dfrac{0.08}{[\text{glucose}]}$

$= +\,3300 + 592 \ln (0.08) - 592 \ln [\text{glucose}]$

$= +\,3300 - 1495 - 592 \ln [\text{glucose}]$

At equilibrium, $\Delta G' = 0$, so $\Delta G' = 1805 - 592 \ln [\text{glucose}] = 0$.

Since $\ln [\text{glucose}] = \dfrac{1805}{592} = 3.049$, [glucose] = **21 M**!

This means that it would require a glucose concentration of 21 M just to bring the reaction to equilibrium; anything over this would render the reaction spontaneous in the direction of phosphorylation. This is impossible; even 2 M glucose would be a thick syrup!

(b) Glucose + ATP $\longrightarrow$ Glucose-6-phosphate + ADP

$\Delta G^{\circ\prime} = +\,3.3\ \text{kcal/mol} - 7.3\ \text{kcal/mol} = -4.0\ \text{kcal/mol}$.

(c) $\Delta G' = \Delta G^{\circ\prime} + RT \ln \dfrac{[\text{glucose - 6 - phosphate}]\,[\text{ADP}]}{[\text{glucose}]\,[\text{ATP}]}$

$= -4000 + (1.987)(298) \ln \dfrac{(0.08 \times 10^{-3})(0.15 \times 10^{-3})}{[\text{glucose}](1.8 \times 10^{-3})}$

$= -4000 + 592 \ln \dfrac{(6.67 \times 10^{-6})}{[\text{glucose}]}$

$= -4000 + 592 \ln (6.67 \times 10^{-6}) - 592 \ln [\text{glucose}]$

$= -4000 - 7055 - 592 \ln [\text{glucose}]$

At equilibrium, $\Delta G' = 0$, so $\Delta G' = -11{,}055 - 592 \ln \text{[glucose]} = 0$.

Since $\ln \text{[glucose]} = \dfrac{-11{,}055}{592} = -18.67$, $\text{[glucose]} = 7.76 \times 10^{-9}$ M!

This means that it would require a glucose concentration of 7.76×10^{-9} M to bring the reaction to equilibrium; any glucose concentration higher than this will render the reaction spontaneous in the direction of phosphorylation. This is physiologically very reasonable, since glucose phosphorylation is thermodynamically feasible as long as the glucose concentration remains above about 0.01 μM.

(d)   From 2.1×10^{1} to 7.76×10^{-9} is about nine orders of magnitude.

(e) $\Delta G' = \Delta G^{\circ\prime} + RT \ln \dfrac{\text{[glucose - 6 - phosphate] [ADP]}}{\text{[glucose] [ATP]}}$

$$= -4000 + 592 \ln \frac{(0.08 \times 10^{-3})(0.15 \times 10^{-3})}{(5.0 \times 10^{-3})(1.8 \times 10^{-3})}$$

$$= -4000 + 592(-6.623)$$

$$= -4000 - 3920 = -7920 = -7.9 \text{ kcal/mol.}$$

5.   (a)   Reaction Gly-6 is balanced by a subsequent step in which pyruvate is reduced to lactate with the net oxidation of NADH to NAD^{+}.

(b)   The starting glucose molecule has no carbon atom higher in oxidation level than an aldehyde and none lower than an alcohol. Lactate, however, has one carbon atom at the carboxylic acid level, representing a more oxidized state than an aldehyde, and one carbon atom at the methyl level, representing a more reduced state than an alcohol. Clearly, one carbon atom has in effect been oxidized and the other reduced during the glycolytic pathway.

(c)

$$
\begin{array}{c}
\underset{\text{Glyceraldehyde-}}{\underset{\text{3-phosphate}}{
\begin{array}{c}
O \\ \| \\ C-H \\ | \\ H-C-OH \\ | \\ H-C-O-\textcircled{P} \\ | \\ H
\end{array}}}
\;+\;
\underset{\text{Pyruvate}}{
\begin{array}{c}
O \\ \| \\ C-O^- \\ | \\ C=O \\ | \\ H-C-H \\ | \\ H
\end{array}}
\;+\;H_2O \longrightarrow
\underset{\underset{\text{glycerate}}{\text{3-Phospho-}}}{
\begin{array}{c}
O \\ \| \\ C-O^- \\ | \\ H-C-OH \\ | \\ H-C-O-\textcircled{P} \\ | \\ H
\end{array}}
\;+\;
\underset{\text{Lactate}}{
\begin{array}{c}
O \\ \| \\ C-O^- \\ | \\ H-C-OH \\ | \\ H-C-H \\ | \\ H
\end{array}}
\;+\;H^+
\end{array}
$$

6. (a) Propionate differs from pyruvate in that its middle carbon atom is at the hydrocarbon level rather than at the carbonyl level. To reduce pyruvate to propionate would therefore require two molecules of NADH, assuming an appropriate sequence of reactions exists. But the stoichiometry of glycolysis provides only one molecule of NADH per molecule of pyruvate, not two. Therefore, all of the pyruvate cannot be reduced to propionate.

   (b) If 50% of the pyruvate molecules are reduced to propionate and the remaining 50% are left as pyruvate, the stoichiometry will come out right. The overall reaction would then be

$$
\underset{\text{Glucose}}{C_6H_{12}O_6} \longrightarrow \underset{\text{Pyruvate}}{C_3H_4O_3} \;+\; \underset{\text{Propionate}}{C_3H_6O_2} \;+\; H_2O
$$

   (c) The pyruvate must be further metabolized by decarboxylation.

7. (a) Galactose + ATP $\longrightarrow$ glucose-6-phosphate + ADP.

   (b) Quite similar, since the net result in both cases is an ATP-linked phosphorylation of a six-carbon sugar, and thermodynamic parameters are independent of route.

   (c)

UDP-galactose $\xrightarrow[\text{NADH,H}^+]{\text{NAD}^+}$ UDP-4-ketoglucose $\xrightarrow[\text{NAD}^+]{\text{NADH,H}^+}$ UDP-glucose

(d)   The inability to metabolize galactose might lead to galactose accumulation in the tissues and blood of an organism that continues to ingest, absorb, and hydrolyze lactose to get the glucose it needs for energy metabolism. Apparently, the elevated level of galactose in the blood is deleterious to brain and lens cells.

8.   (a)   It allows substrate oxidation to proceed without concomitant ATP generation, releasing this step in the glycolytic pathway from its normal sensitivity to, and regulation by, the availability of ADP and P_i.

(b)   Use of arsenate instead of phosphate at step Gly-6 results in spontaneous hydrolysis of the arseno intermediate without conservation of the energy of the bond as ATP. This results in two molecules less of ATP per molecule of glucose, so the energy yield under anaerobic conditions is zero, and the arsenate is therefore fatal.

(c)   Any reaction involving the direct uptake of inorganic phosphate that leads to the generation of a high-energy phosphate bond and the formation of ATP may be subject to uncoupling in this way, provided only that the enzyme will accept arsenate in place of phosphate.

9.   Lactate dehydrogenase (LDH) is an important enzyme in muscle tissue, since it allows muscle cells to carry out fermentation and thereby to continue extracting energy from glucose under anaerobic conditions at times when the supply of oxygen in the blood is not adequate to support aerobic respiration. LDH is therefore a prominent enzyme in the cytoplasm of muscle cells, including heart muscle. A heart attack (or any of several other pathological conditions) involves cell death of heart tissue, with release of cell contents into the blood. Other cytoplasmic enzymes of muscle cells are also released upon tissue damage, including creatine phosphokinase and several transaminases. By assaying the blood of LDH and other enzymes characteristic of muscle cells, clinicians can diagnose the condition, determine the severity of tissue damage, and monitor the progress of treatment.

11
Energy from Chemical Bonds: The Aerobic Mode

1. (a) F   (c) F   (e) T

   (b) T   (d) F   (f) F

2. (a) The two inner carbon atoms (equal likelihood).

   (b) The two carboxyl carbon atoms (equal likelihood).

   (c) The two inner carbon atoms (equal likelihood).

   (d) None (CO_2 evolved in pyruvate dehydrogenase step contains ^{14}C).

3. (a) Carbon atoms 3 or 4.

   (b) Carbon atoms 1, 2, 5 or 6.

   (c) Not possible.

4. Pyruvate + CO_2 + ATP + $H_2O \longrightarrow$ oxaloacetate + ADP + P_i

   Pyruvate + CoA–SH + $NAD^+ \longrightarrow$ acetyl CoA + CO_2 + NADH + H^+

   Acetyl CoA + oxaloacetate + $H_2O \longrightarrow$ citrate + CoA–SH

   Citrate $\longrightarrow$ isocitrate

   Isocitrate + $NAD^+ \longrightarrow$ α-ketoglutarate + CO_2 + NADH + H^+

   α-Ketoglutarate + alanine $\longrightarrow$ glutamate + pyruvate

   ---

   Pyruvate + alanine + ATP + 2 H_2O + 2 $NAD^+ \longrightarrow$

            glutamate + CO_2 + ADP + P_i + 2 NADH + 2 H^+

5. (a) Fluorocitrate has been characterized as the actual poison in the tissues of the animal, and one of the most pronounced of its effects is a blockage of the TCA cycle and a build-up of at least one of the TCA cycle intermediates.

    (b) The suspected blockage point is the conversion of citrate to isocitrate (inhibition of the enzyme aconitase), because (i) it is specifically citrate that accumulates in the tissues of poisoned animals and (ii) fluorocitrate is an analogue of citrate and might be expected to bind and to block the active site of the enzyme that metabolizes citrate in the TCA cycle.

    (c) Fluoroacetate is probably activated to fluoroacetyl CoA and condensed onto oxaloacetate by citrate synthetase to generate fluorocitrate.

    (d) The ingested compound is itself not toxic, but is converted into a lethal metabolite in vivo (that is, a lethal compound is synthesized in vivo from a nonlethal precursor).

6. (a) High levels of ADP mean low levels of ATP, so it is to the advantage of the cell to activate the metabolic pathway responsible for coenzyme reduction, which can in turn give rise to ATP synthesis by electron transport.

    (b) High NADH means adequate reduced coenzyme for the generation of more ATP, so it makes sense to shut down the catabolic machinery of the cell.

    (c) High ATP levels indicate adequate energy supply, so the enzyme responsible for providing the TCA cycle with more acetyl CoA substrate is shut down.

    (d) High citrate levels are indicative of a sufficient supply of acetyl CoA, so the key regulatory enzyme of the pathway leading to pyruvate and acetyl CoA is decreased in activity.

(e) The presence of high levels of acetyl CoA signals the need for more oxaloacetate with which it can condense to initiate the TCA cycle; hence, the logic of activating the enzyme that generates oxaloacetate.

7. (a) Palmitate oxidation to acetyl CoA generates 35 ATP molecules (seven cycles of oxidation) minus the equivalent of 2 molecules of ATP for initial activation, for a net production of 33 molecules of ATP. (Note that to split a 16-carbon chain into eight 2-carbon pieces requires only seven cleavage cycles.)

(b) Each molecule of acetyl CoA yields 12 molecules of ATP, so 8 molecules of acetyl CoA yield 96 molecules of ATP.

(c) 33 + 96 = 129 ATP molecules per palmitate molecule.

(d) Palmitate + 23 O_2 + 129 ADP + 129 P_i $\longrightarrow$

$$16\ CO_2 + 145\ H_2O + 129\ ATP$$

8. (a) For glucose (MW = 180; 36 ATP/glucose):

$$1\ g \times \frac{1\ mol}{180\ g} \times \frac{36\ mol\ ATP}{1\ mol\ glucose} \times \frac{12\ kcal}{mol\ ATP} = 2.4\ kcal/g\ glucose$$

(b) For palmitate (MW = 256; 129 ATP/palmitate; see Problem 7c):

$$1\ g \times \frac{1\ mol}{256\ g} \times \frac{129\ mol\ ATP}{1\ mol\ palmitate} \times \frac{12\ kcal}{mol\ ATP} = 6.0\ kcal/g\ palmitate$$

(c) Fat is 6.0/2.4 = 2.5 times better as a source of energy per gram than carbohydrate. Organisms store energy as fat rather than as carbohydrate because it is more efficient; more calories can be packed into a gram of fat than into a gram of carbohydrate.

(d)   The difference arises because fat is more highly reduced than carbohydrate and hence must be oxidized more to get to the terminal oxidation level of carbon dioxide. More oxidation means more energy, since oxidative reactions are invariably exergonic.

9.   (a)   The reaction will be thermodynamically spontaneous in the direction of oxaloacetate reduction (that is, malate formation) under standard conditions.

(b) $\Delta E_o = E_{o,acceptor} - E_{o,donor} = -0.32 - (-0.166) = -0.154$ V.

$\Delta G^{o\prime} = -nFE_o = (-2)(23,062)(-0.154) = + 7100$ cal/mol.

Reactants are maintained at high concentrations and products at very low concentrations, so that the highly positive $\Delta G^{o\prime}$ value is "overcome" by the term for prevailing concentrations in the expression for $\Delta G'$.

(c)   TCA-3:   feasible.

TCA-6:   not feasible.

The thermodynamic feasibility of reaction TCA-4 is difficult to predict because it requires balancing the oxidative decarboxylation of α-ketoglutarate against both the reduction of NAD^+ and the formation of a thioester bond.

10.   (a)   With antimycin A blocking transport between cytochromes b and c and ferricyanide available to accept electrons from coenzyme Q, electrons will flow from NADH to ferricyanide. By determining the amount of ferricyanide reduced and the amount of ATP generated, we should be able both to verify that ATP generation accompanies electron transport through this segment of the chain and to recognize from the stoichiometry that a single coupling site is involved.

(b)   Independent confirmation comes from the use of succinate as an electron donor in the absence of inhibitors. The generation of only two instead of three ATP molecules per pair of electrons suggests that one coupling site is localized in the segment of the transport chain that is bypassed by electrons from succinate.

(c)   Use succinate as an electron donor, find an artificial acceptor that can receive electrons from cytochrome c, and carry out the transport in the presence of cyanide.

(d)   Use ascorbate as the electron donor, oxygen as the acceptor, and carry out the experiment in the presence of antimycin A.

(e)   All electrons that enter the respiratory chain are passed to molecular oxygen from cytochrome a_3, and cyanide is a potent inhibitor at this step. Accordingly, cyanide blocks all electron flow in respiratory metabolism and is therefore deadly to all aerobic organisms. Cyanide would not be toxic to an anaerobic organism unless it interfered with processes other than the flow of electrons from cytochrome a_3 to oxygen.

11. (a)   If electron transport from FMN to coenzyme Q is blocked, reduced coenzymes cannot be reoxidized, food molecules cannot be oxidized, and no ATP can be synthesized except by glycolysis, which apparently cannot meet energy needs of insects or fish adequately. (This is because higher organisms have no mechanism to excrete lactate directly).

(b)   Rotenone might be expected to inhibit electron transport in other animals as well, because all aerobic cells use very similar electron transport systems, with FMN and coenzyme Q as common features. Therefore, rotenone is potentially poisonous to all animals, unless they cannot absorb the compound (at the level of the intestinal tract, the individual cell, or the mitochondrion) or have mechanisms to detoxify it.

(c)   Plants, like all other eukaryotes, possess mitochondria with an electron transport chain similar to that in animals. Therefore, one might expect rotenone to poison electron transport in plants as well, unless they cannot take up the compound or have a means of detoxifying it.

12. (a)   It is as electrons are passed from an intermediate that transfers both electrons and protons to one that transfers electrons only, or vice versa, that protons are released or taken up by the transport chain, thereby accounting for the pumping of protons on which the electrochemical proton gradient (and therefore the generation of ATP) depends.

    (b)   Iron has a reduction potential in the right range and can be accommodated chemically in organic form as a part of the prosthetic group of electron transfer proteins.

13. (a)   The effect of an uncoupler is to remove the normal control on electron transport, which will then be free to proceed at a rapid, uncontrolled rate. As a result, substrate molecules are oxidized and electrons are transferred rapidly to oxygen at a rate that is insensitive to the availability of ADP or the need for ATP.

    (b)   Lack of transport-driven ATP synthesis will result in low levels of ATP and high levels of ADP, which in turn will activate both the glycolytic pathway and the TCA cycle.

    (c)   The free energy released in electron transport is not conserved as ATP but is quantitatively released as heat, which will tend to raise the temperature of the organism and, in some species, initiate a sweating response.

    (d)   A compound that can facilitate the movement of protons across the membrane unaccompanied by ATP synthesis will tend to "discharge" the proton gradient, leading to the transport of electrons without energy conservation.

    (e)   Since dinitrophenol results in an enhanced rate of cellular respiration, it will cause the body tissues to draw upon and utilize food reserves (rather like exercising without having to move). Its use was discontinued after a death occurred because what was thought to be a "sublethal" dose was in fact lethal.

14. (a) M   (c) M   (e) IM   (g) IM   (i) IM

    (b) IM   (d) IM   (f) IS   (h) IM   (j) IM

15. (a) In; 2   (g) No flux

    (b) In; 6   (h) No flux

    (c) Out; 36   (i) No flux

    (d) In; 36   (j) Out; 6

    (e) No flux   (k) In; 2 pairs per glucose

    (f) No flux   (l) Out; stoichiometry not well established

12

Energy from the Sun: Photosynthesis

1. (a) T   (c) ?   (e) ?

     (b) F   (d) F   (f) T

2. (a) 1   (b) 3, 4   (c) 3, 4, and 5   (d) 1 and 2

3. (a) C_4 plants have as their initial CO_2-fixing enzyme phosphoenol-pyruvate (PEP) carboxylase, which has a higher affinity (lower K_m) for carbon dioxide than does ribulose-1,5-bisphosphate (RuBP) carboxylase, the initial (and only) carbon dioxide-fixing enzyme present in C_3 plants. Carbon dioxide is fixed by PEP carboxylase with high efficiency in the outer (mesophyll) cells and is passed inward in organic form to the bundle sheath cells, where the carbon dioxide is again released, building up an enhanced carbon dioxide concentration in the bundle sheath cells in a way that favors re-fixation by the RuBP carboxylase located there.

    (b) The lower the ambient carbon dioxide concentration, the greater the competitive advantage of a species that has PEP carboxylase instead of RuBP carboxylase, because of the differences in K_m values of the two enzymes for carbon dioxide.

    (c) As carbon dioxide fixation by both species goes on, the carbon dioxide concentration in the sealed container will decrease, giving the C_4 plant a greater and greater competitive advantage because PEP carboxylase has a lower K_m for carbon dioxide. Eventually the carbon dioxide concentration will get so low that the RuBP carboxylase of the C_3 plant cannot function effectively at all, while the PEP carboxylase of the C_4 plant continues to fix carbon dioxide. Beyond this point, the C_3 plant will evolve more carbon dioxide from mitochondrial respiration than it fixes photo-synthetically, resulting in net loss of carbon dioxide and decrease in food reserves. The released carbon dioxide continues to be fixed by the C_4 plant, which therefore gains carbon at the expense of the C_3 plant. This will go on until the C_3 plant has exhausted its carbon reserves and dies.

4. ATP is highly polar and would be difficult to move across membranes; ATP would be a very inefficient way to move energy about, since it has a molecular weight of over 300 and contains only two high-energy phosphate bonds, in contrast to glucose, which has a molecular weight of only 180 and can, upon oxidation, give rise to 36 molecules of ATP elsewhere in the plant; and most importantly, ATP generation stops as soon as the sun goes down, and there is no practical way to store enough of it to meet the ongoing energy needs of the organism during the night.

5. (a) Ribulose-1,5-bisphosphate (RuBP) up, 3-phosphoglycerate (PG) down

    (b) RuBP down, PG up

    (c) RuBP down, PG up

    (d) RuBP down, PG up

6. (a) $6\ CO_2 + 12\ H_2S \longrightarrow C_6H_{12}O_6 + 12\ S + 6\ H_2O$

    (b) The standard reduction potential for the H_2S/S couple is more negative than $+0.4$ V, but that for the H_2O/O_2 couple is not.

    (c) No; the Fe^{2+}/Fe^{3+} couple has a standard reduction potential that is too positive to allow spontaneous transfer of electrons from ferrous ions to bacteriochlorophyll.

7. Flow of energy: photon ──→ excited electron in accessory pigment ──→ excited electron in chlorophyll molecule ──→ excited electron in special chlorophyll molecule at reaction center ──→ reduced electron acceptor ──→ reduced intermediates in electron transport chain ──→ electrochemical proton gradient ──→ ATP ──→ high-energy phosphate bond of 1,3-bisphosphoglycerate ──→ glyceraldehyde-3-phosphate ──→ other reduced intermediates ──→ sugar.

8. Although both the mint sprig and the mouse require air, they do not depend upon it for the same thing; the mint sprig needs carbon dioxide, whereas the mouse requires oxygen. In fact, the photosynthetic activity of the mint sprig enhances the oxygen supply of the air, rendering it "not at all inconvenient" to the mouse. The observation was important in establishing that whatever a photosynthetic organism required from air, it was obviously not the same thing that a respiring animal needed, but was in fact complementary to it.

9. (a) Valid   (b) Valid   (c) Invalid   (d) Valid

10. (a) Higher (more noncyclic activity is required to meet the demand for extra NADPH).

    (b) Lower (more cyclic activity to meet the demand for extra ATP).

    (c) Lower (more cyclic activity to meet the demand for extra ATP).

    (d) Lower (noncyclic activity is inhibited).

    (e) Higher (cyclic activity is inhibited and noncyclic flow is enhanced by the presence of ferricyanide as an artificial acceptor).

11. (a) $7 \text{ metric ton} \times \dfrac{1000 \text{ kg}}{\text{metric ton}} \times \dfrac{1000 \text{ g}}{\text{kg}} \times \dfrac{1 \text{ mol sucrose}}{342 \text{ g}}$

$\times \dfrac{12 \text{ mol CO}_2}{1 \text{ mol sucrose}} \times \dfrac{10 \text{ einsteins}}{1 \text{ mol CO}_2} = 2.46 \times 10^6 \text{ einsteins.}$

(b) $\dfrac{10,000 \text{ m}^2}{1 \text{ hectare}} \times \dfrac{10,000 \text{ cm}^2}{1 \text{ m}^2} \times \dfrac{108 \text{ days}}{\text{season}} \times \dfrac{15 \text{ h}}{\text{day}} \times \dfrac{3600 \text{ sec}}{1 \text{ h}}$

$\times \dfrac{1.4 \text{ cal}}{\text{cm}^2 \cdot \text{sec}} \times \dfrac{1 \text{ einstein}}{55,000 \text{ cal}} = 1.48 \times 10^{10} \text{ einsteins.}$

(c) $\text{Efficiency} = \dfrac{2.46 \times 10^6}{1.48 \times 10^{10}} \times 100\% = 1.7 \times 10^{-4} = 0.017\%.$

(d) Not all light is of the right wavelength; not all of the surface is covered by leaves throughout the growing season; the photosystems become saturated with light far below the maximum intensity of sunlight; and other factors (such as temperature and availability of water) may limit growth more than the availability of light.

12. (a) Stroma.

(b) The smaller particles in the thylakoid membrane (photosystem I).

(c) Both kinds of particles in the thylakoid membrane.

(d) CF_1 (ATP synthase) particles protruding outward from the thylakoid membrane.

(e) Across the thylakoid membrane from the cytoplasm into the intrathylakoid space.

(f) The smaller particles in the thylakoid membrane (photosystem I).

(g) Stroma.

(h) Within the thylakoid membrane.

13. (a) Inward across the outer and inner chloroplast membranes.

    (b) Across the thylakoid membrane from stroma into intrathylakoid space.

    (c), (d), (e), (f), and (g): No flux across the membrane.

14. (a) The products of fatty acid oxidation and their fate are:

    Acetyl CoA: converted via succinate to hexoses.

    $FADH_2$: reoxidized to FAD by oxidase activity, consuming oxygen and generating hydrogen peroxide.

    NADH: cannot be reoxidized within glyoxysome; must be passed to outside (see part f below).

    (b) One molecule of palmitate gives rise to eight molecules of acetyl CoA and hence to four molecules of succinate. This requires seven cycles of β oxidation, so seven molecules each of $FADH_2$ and NADH are formed.

    (c) Four molecules of succinate give rise to four molecules of phosphoenolpyruvate and hence to two hexose molecules. The sequence from succinate to oxaloacetate is part of the TCA cycle and occurs in the mitochondrion; the remainder of the pathway takes place in the cytoplasm. Other products include: the $FADH_2$ and NADH formed by mitochondrial oxidation of succinate and malate, respectively; the carbon dioxide produced in the conversion of oxaloacetate to phosphoenolpyruvate; and the NAD^+ generated as glycerate-3-phosphate is reduced to glyceraldehyde-3-phosphate in the reversal of the glycolytic pathway necessary to convert phosphoenolpyruvate to hexoses.

    (d) Since a molecule of palmitate gives rise to two hexose molecules, the end product will be a single sucrose molecule. Of the original 16 carbon atoms in palmitate, 12 appear in sucrose. The other four carbon atoms are evolved as carbon dioxide when oxaloacetate is converted to phosphoenolpyruvate.

(e) Since one molecule of palmitate ($C_{16}H_{32}O_2$; MW = 256) gives rise to one molecule of sucrose ($C_{12}H_{22}O_{11}$; MW = 342), each gram of stored palmitate will yield about 1.34 (342/256) g of sucrose.

(f) NADH moves outward for reoxidation outside the glyoxysome, NAD^+ moves inward to allow continued oxidation of fat, and oxygen moves inward to provide for direct oxidation of $FADH_2$. The ATP needed for activation of fatty acids (reaction FA-1) must be supplied from outside the organelle, so ATP moves inward also, presumably accompanied by exchange of ADP outward.

13

The Flow of Information: DNA, Chromosomes, and the Nucleus

1. (a) I   (c) A   (e) N   (g) B

    (b) B   (d) B   (f) I   (h) N

2. (a) It made their experiment possible, since it allowed them to assay various fractions of S cells for transforming activity in vitro.

    (b) This was the very hypothesis their experiment was designed to test; all that was left to do was to devise a means of identifying the nucleic acid if it did indeed "flow into the cell."

    (c) The concept of hydrogen-bonded base pairing between pyrimidines and purines on opposite strands turned out to be a vital clue to the double-stranded structure of the DNA molecule.

    (d) By showing that a virus was indeed like a "little hypodermic needle" capable of injecting its nucleic acid into the cell, Hershey and Chase were able to explain Anderson and Herriott's observation in terms of an osmotic shock that causes the viruses to empty their nucleic acid contents into the medium.

3. (a) T   (b) T   (c) F   (d) T   (e) F

4. When the DNA of Problem 3 is labeled at the 5' end with ^{32}P (represented below as *P) and then treated with dimethyl sulfate-piperidine to delete guanines, the following radioactive fragments will be formed, depending on which guanine happens to be deleted in a specific DNA molecule:

Fragment 1:   *P-TC   Fragment 3:   *P-TCGCGATATC

Fragment 2:   *P-TCGC   Fragment 4:   *P-TCGCGATATCGC

When subjected to gel electrophoresis, these fragments will migrate down the gel at a rate that is inversely related to their length, such that the smallest fragment will move the farthest, and the largest fragment will move the least. (In addition, some of the original radioactive DNA molecules are likely to remain intact and will move even more slowly than the largest fragments.) When the gel is exposed to a photographic film, the band pattern shown in Figure A13-1 will be observed.

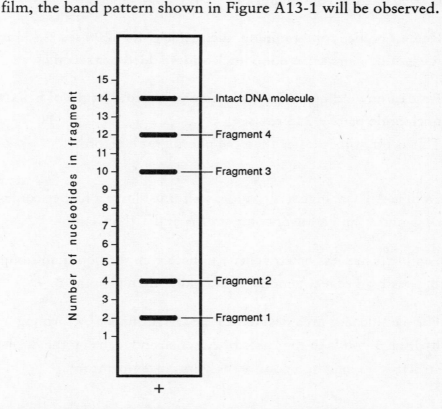

Figure A13 -1.   Banding pattern for the DNA fragments of Problem 4.

5. (a) $(4 \times 10^6$ nucleotide pairs$)(0.34$ nm/nucleotide pair$) =$

1.36×10^6 nm = **1360 μm.**

(b) Radius = circumference/2π = 1360/6.28 = **216 μm.**

A circle of this radius could not possibly be fully extended in a cell that is only about 2 μm long.

(c) Molecular weight: $(4 \times 10^6$ np$)(660$ g/mol$)=$ **2.64×10^9 daltons.**

Weight: $(2.64 \times 10^9$ g/mol$)(1$ mol/6×10^{23} molecules$) =$ **4.4×10^{-15} g.**

(d) Volume of DNA: $(4.4 \times 10^{-15}$ g$)(1$ cm^3/1.7 g$) =$ **2.6×10^{-15} cm^3.**

Volume of cell: $\pi r^2 h = (3.14)(0.5)^2(2) = 1.57$ μm$^3 =$ **1.57×10^{-12} cm^3.**

So DNA occupies $(2.6 \times 10^{-15})/(1.57 \times 10^{-12}$ cm$^3) = 0.0016 =$ **0.16% of total volume of cell.**

(e) Rate of replication: 4 million nucleotide pairs/1200 sec = 3333 nucleotide pairs/sec = **6666 nucleotides added per second.**

(f) Rate of unwinding: 3333 nucleotide pairs/sec $\times$ 1 turn of helix/10 nucleotide pairs = **333 rev/sec.**
This is 600 times faster than the phonograph record.

6. (a) Sample A has a higher T_m value, so it must have a higher content of G and C and a lower content of A and T than sample B.

(b) Sample A has less heterogeneity in base composition than sample B, possibly because genome A is smaller than genome B.

(c) Formamide and urea destabilize the DNA duplex by forming hydrogen bonds to the bases of either strand, causing the duplex structure to melt at a considerably lower temperature.

7. (a) Unlike most membranes, the nuclear envelope appears to be freely permeable to a polar organic molecule.

(b) The diameter of the channel in the nuclear pore is at least 5.5 nm, but not as great as 15 nm.

(c) The pore probably contains both RNA and protein, either as an integral structural part of the pore complex or as a particle caught in transit.

(d) The nuclear membranes and the endoplasmic reticulum are likely to have a common origin.

(e) Ribosomal proteins must pass inward from the cytoplasm to the nucleus at a rate adequate to sustain ribosomal subunit assembly; ribosomal subunits must move outward from the nucleus to the cytoplasm at a rate commensurate with their rate of assembly.

(f) Nucleoli apparently form at secondary constrictions.

(g) Nucleoli are probably responsible for rRNA synthesis.

(h) The integrity of the nucleus does not seem to depend entirely on the envelope.

8. (a) Total number of pores = (50 pores/μm^2)(200,000 μm^2) = **10^7 pores.**

(b) Since each square micrometer contains 50 pores and $50 \approx 7^2$, each square micrometer contains a 7×7 matrix of pores. Pores are therefore 1000/7 = 140 nm apart from center to center.

(c) The area of one pore can be calculated as $A = \pi r^2 = (3.14)(37.5)^2 = 4400$ nm^2. The area of 50 pores is therefore $4400 \times 50 = $ **220,000 nm^2.** Thus, pores represent 22% of the surface area. This is a high value.

(d) Oocyte development apparently requires a great deal of nucleocytoplasmic exchange, probably because the oocyte is stockpiling materials that the developing embryo will need after fertilization.

14

The Cell Cycle, DNA Replication, and Mitosis

1. (a) S   (d) G1   (g) G1, G2, S

   (b) M, C   (e) C

   (c) M   (f) G1, G2, S, possibly C

2. (a) G-1 has the 2C amount of DNA; G-2 has the 4C amount.

   (b) DNA is synthesized during S; little or no synthesis occurs during G-1.

   (c) Chromosomes are in an extended form during G-2, but in a condensed form during mitosis.

   (d) Chromosomes separate during mitosis; cytoplasm is partitioned into two daughter cells during cytokinesis.

3. (a) 25% hybrid ($^{14}N/^{15}N$) DNA, 75% ^{14}N DNA.

   (b) 12.5% hybrid DNA, 87.5% ^{14}N DNA.

   (c) 25% ^{15}N DNA, 75% hybrid DNA.

4. Your sketch should look like Figure 14-14, with single-stranded DNA identified as the template, Okazaki fragments identified as nascent pieces of DNA on the lagging strand, short segments of RNA on the Okazaki fragments identified as primers, and helix-destabilizing proteins bound to the single-stranded DNA near the point at which the

fork is opening. In addition, your sketch should include the following enzymes:

(a) Helicase: unwinds double-stranded DNA.

(b) Gyrase: nicks DNA ahead of the replication fork to relax the supercoiling induced in the DNA by helicase activity.

(c) Primase: synthesizes the short segments of RNA used as primer for DNA synthesis.

(d) DNA polymerase III: elongates the growing segment of DNA.

(e) DNA polymerase I: removes RNA primer and replaces it with deoxyribonucleotides.

(f) Ligase: links discontinuous fragments of DNA covalently into a continuous strand.

5. (a) R; in *E. coli*, exonucleolytic activity is needed for RNA primer removal and for "editing."

(b) R; in *E. coli*, synthesis is continuous on one strand and discontinuous (with Okazaki fragments) on the other.

(c) NB; the degree of sequence reiteration is not likely to have any effect on the mode of replication of the DNA.

(d) S; Okazaki fragments have short RNA (ribose) segments as well as longer DNA (deoxyribose) segments.

(e) R; this finding is not consistent with the semiconservative mode of replication found in *E. coli* DNA.

6. (a) DP   (d) DF   (g) DF   (j) None

(b) DF   (e) None   (h) DP, DA, DF

(c) DP   (f) DA   (i) DP, DA, DF

7. (a) Uracil arises from cytosine, hypoxanthine from adenine, and xanthine from guanine.

   (b) Thymine does not contain an amino group and therefore cannot be deaminated.

   (c) The excision process would not be able to distinguish between a base that has been generated by deamination and a naturally occurring base.

   (d) The base 5-methylcytosine yields thymine upon deamination, and there is no mechanism to detect and excise thymine bases. Deamination of 5-methylcytosine will therefore go unrepaired, and every such deamination will convert a C into a T.

8. (a) Pyrimidine dimers cannot fit into a double helix. They therefore block replication and transcription.

   (b) The defective strand will be cleaved at every site of dimer formation as part of the repair process. If the DNA is extracted and denatured before repair is completed, each such site will be a point of breakage of single-stranded DNA.

   (c) The absence of any effect of ultraviolet light on the size of single-stranded DNA suggests that no cleavage has occurred at sites of dimer formation, indicating a possible defect in the nuclease responsible for this cleavage.

   (d) Since the defect is rare and appears to be due to a nonfunctional gene, it is almost certainly a recessive trait.

   (e) Stay out of the sun!

9. (a) Mitotic index = (50 + 20 + 10 + 20)/1000 = 0.1 = **10%**.

   (b) 5% of time in prophase, 2% in metaphase, 1% in anaphase, 2% in telophase, 40% in G-1, 30% in S, and 20% in G-2.

(c) Since G-2 is 20% of the cell cycle and lasts 4 hours, the whole cycle must be 4/0.2 = **20 hours.**

(d) 1 h in prophase, 0.4 h each in metaphase and telophase, 0.2 h in anaphase, 8 h in G-1, 6 h in S, and 4 h in G-2.

(e) The first appearance of label in prophase nuclei would have to be observed.

(f) About 30%, since all cells in S phase will incorporate label immediately, and on the average about 30% of the cells in the culture should be in S phase at any one time.

10. (a) Embryonic cells: mitosis without cytokinesis (multinucleate cells)

Liver cells: DNA replication and chromosome separation without subsequent partition into daughter nuclei (polyploid cells)

Salivary gland cells: DNA replication without subsequent separation of chromatids (polytene chromosomes)

(b) Embryonic cells; many insect and amphibian embryos undergo extended periods of rapid division unaccompanied by a growth phase because the egg has sufficient cytoplasmic mass to sustain many rounds of division.

(c) No; the salivary gland is a differentiated tissue, and differentiated tissues generally do not continue to divide. Moreover, the chromosomes contain about 1000 times more DNA than usual, indicating that substantial DNA replication has already occurred, unaccompanied by cell division or even mitotic segregation.

(d) Surgical removal of a portion of the liver may induce resumption of cell division.

11. (a) T   (c) F   (e) T

(b) F   (d) NP   (f) F

12. (a) Inconsistent. If the microtubules disassembled at the spindle pole, both the dark (photobleached) band and the chromosomes would move at the same rate. If such were the case, the chromosomes would remain the same distance from the dark band, rather than appearing to move toward it.

    (b) Consistent. (This statement appears to be the correct explanation for chromosome movement.)

    (c) Inconsistent. Again, if microtubules were contracting in some way, the chromosomes and the photobleached band would move at the same rate.

    (d) Consistent. (This explanation does not appear to be correct, however. If a "motor" protein were moving the chromosomes, the spindle fibers should remain behind the moving chromosomes, but this does not appear to be the case.)

15

Sexual Reproduction, Meiosis, and Genetic Variability

1. (a) S   (b) A   (c) N   (d) S   (e) N

2. (a) 2n = 4

   (b) Correct order: F, E, D, B, A, C.

   A: metaphase II

   B: telophase I

   C: anaphase II

   D: metaphase I

   E: diplotene of meiosis I

   F: leptotene of meiosis I

   (c) Between metaphase I (D) and telophase I (B).

   (d) Between leptotene (F) and diplotene (E) of meiosis I.

3. (a) Homologous chromosomes pair at meiotic metaphase I, but not at mitotic metaphase.

   (b) Meiotic metaphase II has only one-half as many chromosomes on the metaphase plate as does mitotic metaphase.

(c) Compared to metaphase I, metaphase II has only one-half as many chromosomes on the metaphase plate and no paired bivalents.

(d) Mitotic telophase yields two diploid daughter nuclei; meiotic telophase II yields four haploid daughter nuclei.

(e) Chromosomes of each bivalent have begun to separate and chiasmata have become visible by diplotene, but not by pachytene.

4. (a) Yes. Each centromere is duplicated and one of each is passed to each of the daughter cells resulting from mitosis. All somatic cells arise from the zygote by mitosis, and the products of mitosis are always equal.

    (b) No. Centromeres separate at the first meiotic division, and independent assortment will randomize the segregation of maternal and paternal centromeres. Each first-division daughter cell will receive a maternal or a paternal centromere for each homologous pair.

5. (a) 2X   (c) 2X   (e) 4X   (g) 2X

    (b) $\frac{1}{2}$X   (d) $\frac{1}{2}$X   (f) 2X

6. (a) The four possible genotypes of the offspring, in their respective ratios, are 1 YY, 2 Yy, and 1 yy. Since all YY and Yy offspring will be yellow and only yy offspring will be green, offspring with yellow and green seeds will appear in the ratio 3:1.

    (b) Both parents are heterozygous for both factors, so the number of gametes is 2^2, or 4. A 4×4 matrix has 16 outcomes for offspring, although not all of the 16 are different. In general, the number of different gametes is 2^n, where n is the number of heterozygous allelic pairs.

(c) The Punnett square assumes that the frequency of occurrence of a given genotype among the boxes represents the frequency of occurrence of that genotype among the progeny of the genetic cross represented by the Punnett square. This assumption is only correct if all possible combinations are equally likely, which in turn depends on independent assortment of the alleles for seed color and seed shape.

(d) The completed matrix will have 16 genotypes, nine of which are different from one another. In their respective ratios, these are 1 YYRR, 2 YYRr, 1 YYrr, 2 YyRR, 4 YyRr, 2 Yyrr, 1 yyRR, 2 yyRr, and 1 yyrr.

(e) The phenotypes are yellow and round, yellow and wrinkled, green and round, and green and wrinkled. These occur in the ratio 9:3:3:1.

7. (a) The eight possible homozygous genotypes are AALLRR, AALLrr, AAllRR, aaLLRR, AAllrr, aaLLrr, aallRR, and aallrr. Each of these occurs with a frequency of 1 in 64, so the combined frequency of all homozygous offspring should be 8 in 64, or 1 in 8.

(b) In this case, there are only four homozygous genotypes (AALLRR, AAllRR, aaLLRR, and aallRR), but each occurs with a frequency of 2 in 64, so the overall frequency of homozygous genotypes will still be 8 in 64, or 1 in 8.

(c) One of the parents must be homozygous dominant and the other homozygous recessive for at least one of the three traits under study.

(d) If two of the genes were on the same chromosome, they would probably not segregate independently, and we could not then assume that all possible progeny genotypes would occur with equal probability.

16

From Genes to Proteins: The Genetic Code and Protein Synthesis

1. (a) Degenerate: A given amino acid can be specified by more than one codon; G

   (b) Unambiguous: A given codon specifies a single amino acid; G, M (Exception: In cells containing suppressor tRNA, the code is ambiguous.)

   (c) Triplet: There are three nucleotides in every codon; G

   (d) Universal: The same code is used by all organisms; G (except for mitochondria); M

   (e) Nonoverlapping: Adjacent codons do not share common nucleotides; G, M

2. (a) NO   (d) NO   (g) NO

   (b) OK   (e) OK   (h) OK

   (c) NO   (f) OK   (i) OK

3. (a) Single-base substitution:

   (i) One (or possibly no) amino acid change.

   (ii) Change of 1, 2, or 3 amino acids (depending on wobble and degeneracy of code).

(b) Single-base deletion:

    (i)   Frameshift, with likely change of all downstream amino acids.

    (ii) Loss of one amino acid, change of none, 1, or 2 others.

(c) Deletion of three consecutive bases:

    (i)   Loss of one amino acid, possible change in another, depending on reading frame.

    (ii) Loss of three amino acids, change of none, 1, or 2 others.

4. (a) 5'   UUAAU<u>AUG</u>UGCUACUUCGAACACUGUCCCAAAGGUU<u>AG</u>U<u>AA</u>UU   3'

       3'   AAUUAUACACGAUGAAGCUUGUGACAGGGUUUCCAAUCAUUAA   5'

(b) Only the first of the two RNA sequences has an initiation codon (AUG, underscored above) in the correct (5'→3') direction, so it must be the messenger. Notice also that the same RNA sequence has two terminator codons in the correct reading frame near the 3' end (UAG, UAA, also underscored). The amino acid sequence of vasopressin is therefore:

AUG UGC UAC UUC GAA CAC UGU CCC AAA GGU UAG UAA

Met-Cys-Tyr-Phe-Glu-His-Cys-Pro-Lys-Gly-(term.)

(0) 1   2   3   4   5   6   7   8   9

(c) The methionine at the N-terminal end is apparently cleaved from the nascent peptide to generate the mature polypeptide.

(d) Oxytocin differs from vasopressin in positions 3 (Ile for Phe), 5 (Asp for His), and 8 (Leu for Lys). The first two can be accomplished by one-base changes in the mRNA (AUC instead of UUC, GAC instead of CAC), but the latter requires a two-base change (UUA or CUA instead of AAA). The most conservative DNA sequence would therefore be:

3'   AATTATACACGATGTAGCTTCTGACAGGGAATCCAATCATTAA   5'

5. (a) T   (b) F   (c) F   (d) T   (e) T

6. (a) B, I   (e) B, I

    (b) B, I, II, III   (f) none

    (c) I, II, III   (g) B, I, II, III

    (d) B   (h) II

7. (a) rRNA: Cleavage of large precursor; degradation of unconserved portions.

    tRNA: Removal of leader sequence at 5' end; addition of CCA sequence at 3' end; methylation of bases; removal of introns.

    mRNA: Elimination of introns; addition of poly-A tail on 3' end; capping of 5' end.

    (b) rRNA: Generation of several molecules from one large precursor.

    tRNA: Addition of CCA sequence at 3' end.

    mRNA: Capping of 5' end.

    (c) rRNA: Degradation of unconserved sequences.

    tRNA: Addition of CCA sequence.

    mRNA: None in common.

8. (a) Rifamycin interferes with initiation because it prevents formation of the first phosphodiester bond.

    (b) Actinomycin D interrupts elongation because it blocks the DNA.

    (c) Actinomycin D, because rifamycin is specific for bacterial polymerase.

    (d) Rifamycin, because it inhibits initiation but does not block elongation.

9. (a) Because it looks like an aminoacyl tRNA and contains an α amino acid, puromycin can bind to the ribosome and form a peptide bond with the carboxyl group of the growing peptide chain. The product, peptidyl puromycin, then dissociates from the ribosomes.

   (b) To the carboxyl end, because it has an amino group.

   (c) To the A site, because it resembles an aminoacyl rRNA.

   (d) Equally effective.

10. (a) Codons: UUU, UUC, UCU, CUU, CCU, CUC, UCC, CCC.

        Amino acids: phenylalanine (UUU, UUC), serine (UCU, UCC), leucine (CUU, CUC), proline (CCU, CCC).

   (b) Incubation A: 8 of each.

        Incubation B: 27 UUU; 9 each of UUC, UCU, and CUU; 3 each of CCU, CUC, and UCC; and 1 CCC.

   (c) Incubation A: 16 of each.

        Incubation B: 36 phenylalanine; 12 each of serine and leucine; 4 proline.

   (d) It is possible to distinguish between classes of codons that differ in their base composition and to determine the base compositions of the codons that code for various amino acids.

   (e) Yes, the data of parts (b) and (c) for incubation B should allow this to be deduced.

   (f) No.

   (g) Synthesize the relevant codons and test them in the rRNA-binding assay of Nirenberg.

11. Sense RNA is synthesized very readily by using RNA polymerase to transcribe a coding DNA template that contains a promoter region and terminal signal. The RNA polymerase attaches to the promoter, separates the double helix, and transcribes sense RNA in the $5' \rightarrow 3'$ direction. Antisense RNA is more difficult to make, since RNA polymerase must use the opposite (noncoding) DNA strand as a template and would therefore need to read the DNA in the wrong (i.e., $3' \rightarrow 5'$) direction. To get around this problem, an additional step is necessary, in which a restriction enzyme is used to cleave the DNA molecule in the region where it is desired to make an antisense RNA. The restriction fragments are then dissociated from the larger portion of DNA (a plasmid, typically), followed by annealing back into the plasmid. Because the fragments have complementary ("sticky") ends, about half of the fragments anneal in the original direction, and half anneal so that the antisense strand can be read by RNA polymerase. The strands are then treated with DNA ligase to produce the final template, which can then be used for transcription into RNA by RNA polymerase.

12. (a)   Signal sequences will consist primarily of hydrophobic amino acids.

    (b)   Proteins destined for the same membranous compartment may be synthesized with the same or very similar signal sequences on the amino end of the molecule.

    (c)   The signal peptide must have a common sequence that can be recognized by the signal peptidase.

13. (a)   Because it is an analogue of UDP-N-acetylglucosamine, tunicamycin blocks the addition of N-acetylglucosamine to oligosaccharide chains.

    (b)   Tunicamycin is more effective in blocking core glycosylation because the first step in the formation of the core oligosaccharide is the addition of N-acetylglucosamine to dolichol phosphate.

(c) Treat secretory cells with tunicamycin and observe the extent to which proteins known normally to undergo both core and terminal glycosylation do or do not have oligosaccharide side chains added to them in the presence of the antibiotic.

14. The insulin that appears in the blood immediately after a meal preexists in the pancreatic cells and is being released from secretory granules in the cell. The time lag between injection with labeled leucine and the appearance of radioactivity in insulin results in part from the time needed for the labeled leucine to reach and be absorbed by the pancreatic cells and in part from the time needed to synthesize, modify, concentrate, and package the insulin for secretion.

15. (a) G-protein is synthesized by membrane-bound ribosomes on the rough ER, is transported to the Golgi for processing and packaging, and is carried, probably by coated vesicles, to the plasma membrane.

(b) The amino end of the protein is exposed to the interior (luminal side) of the rough ER, the Golgi, and the vesicles in which it is transported and to the exterior side of the plasma membrane. The carboxyl end is exposed to the cytoplasmic side of all membranes in which the protein is embedded.

(c) Galactose is attached to the G-protein in the Golgi because this is the first structure in the cell in which ^{3}H-labeled G-protein appears when ^{3}H-labeled galactose is used. Labeled galactose appears in the Golgi fraction before labeled amino acids because the sugar is added directly to proteins already present in the Golgi, whereas proteins are synthesized on the rough ER, so that proteins containing labeled amino acids can only be found in the Golgi after they have been synthesized on the rough ER and transported to the Golgi.

17

The Regulation of Gene Expression

1. Activity of the z Gene:

	– Inducer	+ Inducer	Explanation
(a)	–	+	Wild–type; inducible system
(b)	–	–	Superrepressor cannot recognize inducer
(c)	+	+	Operator cannot bind repressor
(d)	+	+	Repressor not made; system always "on"
(e)	+	+	Operator cannot bind repressor
(f)	–	–	Promoter cannot bind RNA polymerase
(g)	–	–	Superrepressor cannot recognize inducer; binds to both operators
(h)	–	–	Same as for part (g)
(i)	–	+	Normal repressor binds to both normal operators
(j)	–	–	Glucose reduces CAP binding

2. Gene a is the operator o; its absence results in constitutive synthesis (line 1), even when a normal allele is present (line 7). Gene b is the structural gene z; its absence results in no synthesis (line 3).

Gene c is the repressor i; its absence results in constitutive synthesis (line 2) unless a normal allele is also present (lines 5 and 6).

3. (a) Genes a and b are structural genes, probably coding for enzymes involved in the pathway for ethanol synthesis. Gene c is the operator, and d is the repressor.

    (b) (i)   No synthesis of a, inducible for b.

         (ii) Inducible for both enzymes.

         (iii) Inducible for both enzymes.

         (iv) Inducible for a, constitutive for b.

4. (a) C   (c) X   (e) C   (g) C   (i) I

    (b) C   (d) X   (f) I   (h) C   (j) X

5. (a) their mode of action is to terminate transcription prematurely to varying extents under different conditions of concentration.

    (b) signals termination, causing dissociation of the RNA polymerase from the template.

    (c) the transcriptional activity of an RNA polymerase molecule is influenced by changes in mRNA structure induced by a ribosome that is translating that mRNA.

    (d) it ensures that the ribosome does indeed "tailgate" the RNA polymerase, allowing the necessary proximity of the ribosome to the RNA polymerase molecule.

    (e) has amino acid X present at more than one position.

    (f) the model requires sensitivity of transcription to aminoacyl tRNA availability and therefore to the presence of a specific amino acid in the leader peptide.

(g) the close coupling of transcription and translation on which the model depends is not possible in eukaryotes because of the nuclear membrane.

6. (a) B, G   (c) N   (e) L   (g) L

   (b) L   (d) G   (f) N   (h) N

7. (a) The number of complementation groups is the same as the total number of chromosomal bands, suggesting that each band corresponds to a single gene. The number of proteins would therefore be 5000, which seems barely adequate to code for all the structural and functional needs of a eukaryotic organism, given that *E. coli* is estimated to have about 2500 to 3000 proteins.

   (b) $(2 \times 10^8)(0.75)/5000 = 30,000$ nucleotide pairs per band.

   (c) $50,000/110 = 455$ amino acids per protein.

       $455 \times 3 = 1365$ nucleotide pairs per protein.

       $1365/30,000 = 0.045 = 4.5\%$.

   (d) The discrepancy could result either from DNA that is present in the band but is not transcribed or from RNA that is transcribed but does not actually code for anything. Since RNA transcripts in eukaryotic nuclei (including those of *Drosophila* salivary glands) are notoriously larger than cytoplasmic messengers, the latter explanation is probably correct.

8. (a) $(500)(13,000)/(2.7 \times 10^9) = 0.0024 = 0.24\%$.

   (b) $(2.7 \times 10^9) + (5 \times 10^5)(13,000) = 9.2 \times 10^9$ nucleotide pairs per amplified haploid genome.

       Ribosomal genes represent $6.5/9.2 \times 100\%$, or 70%, of the amplified genome!

(c) Since amplification is 1000-fold, the unamplified genome would presumably require 1000 times longer to synthesize the required number of ribosomes. Therefore 2000 months, or about 167 years, would be needed (which would make female frogs *very* old mothers!).

(d) When the desired gene product is an RNA, all the molecules the cell needs must be transcribed directly from the DNA. When the desired product is a protein, its mRNA can be translated repeatedly, so there is an additional stage of amplification at the mRNA level.

9. (a) The total mRNA population of one tissue could be radioactively labeled and hybridized against total mouse DNA (or, for greater sensitivity, against the unique-sequence component of the mouse genome) in the presence and absence of mRNA from the other tissue. The extent to which radioactive RNA from one tissue could be competed out by nonradioactive species from the other tissue measures the proportion of common sequences.

(b) Selective processing of nuclear transcripts and/or selective translocation of RNA to the cytoplasm could also explain the data. To distinguish between these possibilities, nuclear RNA preparations from the two tissues could be tested for the presence or absence of common sequences by the same hybridization techniques used with the cytoplasmic RNA.

(c) Translational control would seem likely.

10. (a) C   (c) C   (e) C   (g) C

    (b) X   (d) I   (f) I   (h) X

18
Cytoskeletal Structure and Function

1. (a) MF (c) MT (e) IF (g) IF   (i) MF, MT

    (b) MT (d) MF (f) N   (h) MF, MT, IF (j) MT, IF

2. (a) T

    (b) F; hydrolysis of the ATP bound to actin and the GTP bound to tubulin usually occurs during monomer polymerization, but the polymerization process still occurs even if the ATP or GTP is replaced by a nonhydrolyzable analogue.

    (c) F; the statement is true for microtubules and microfilaments, but there is no evidence for dynamic assembly and disassembly of intermediate filaments.

    (d) F; cytochalasin B prevents actin polymerization and therefore inhibits cytokinesis but not chromosome movement, since actin is involved in the former but not the latter.

    (e) M; the statement is true for a blue-green algal cell, but not for other kinds of algae, since blue-green algae are prokaryotes, but all other algae are eukaryotes.

    (f) T

    (g) F; most eukaryotic cells contain nuclear lamins in the nucleus and at least one kind of intermediate filament protein in the cytoplasm.

(h) M; the statement is true if the monomer concentration is above the critical point, but false otherwise.

3. (a) (13.5 monomers/72 nm)(42,000 daltons/monomer)(100 nm)(2 strands) = 1.58×10^6 daltons.

   (b) (1 dimer/8 nm)(2 monomers/dimer)(50,000 daltons/monomer)(100 nm)(13 protofilaments) = 1.63×10^7 daltons.

   (c) Microtubules are $(1.63 \times 10^7)/(1.58 \times 10^6)$ = 10.3 times heavier than microfilaments per unit length.

   (d) (1 protofilament/48 nm)(4 polypeptides/protofilament)(53,000 daltons/polypeptide)(8 protofilaments)(100 nm)= 3.53×10^6 daltons.

   (e) A vimentin filament is $(3.53 \times 10^6)/(1.58 \times 10^6)$ = 2.23 times heavier than a microfilament and only $(3.53 \times 10^6)/(1.63 \times 10^7)$ = 0.217 times as heavy as a microtubule, on a per unit length basis.

4. (a) Tubulin   (d) Actin

   (b) Keratin   (e) NF protein

   (c) Desmin

5. (a) F-actin and myosin function together to convert the energy of ATP into movement, without the involvement of any other proteins or structures.

   (b) The centrosome serves as a microtubule-organizing center in vivo, and all of the microtubules radiating from the centrosome apparently have the same polarity.

   (c) Although each kind of muscle cell has actin molecules that are specific for that muscle type, the differences between the several species of actin are apparently too small to have any discernible effect on actin polymerization, structure, or function.

(d) The polymerization of microtubules into the mitotic spindle is effectively regulated by the presence or absence of the centrosome.

6. (a) A   (c) B and C   (e) N

    (b) C   (d) B   (f) C

7. (a) It would be reasonable to postulate that the G-actin is kept from polymerization prior to contact with an egg cell by being complexed one-to-one with profilin. The profilin-actin complex, unlike free G-actin, would not be able to initiate polymerization. To test this hypothesis, profilin could be isolated and added to a mixture of G-actin monomers in vitro to determine whether the profilin can bind to actin and prevent the initiation of microfilament assembly under conditions that would normally be optimal for polymerization.

    (b) Since the pH rise precedes actin polymerization, it would be reasonable to postulate that the change in pH destabilizes the profilin-actin complex, causing dissociation into free actin, which can then assemble into F-actin. To test this hypothesis, one could vary the pH of the mixture of profilin and actin and determine whether profilin-actin complexes are present at high but not low pH. One could also examine the effects of pH on actin polymerization in the presence of profilin in vitro.

8. (a) One way to test the involvement of microtubules would be to treat the cells with colchicine, which should disrupt microtubule-mediated processes. The ER might be expected to contract and lose its shape in colchicine-treated cells. Alternatively, immunofluorescence microscopy could be used to determine whether microtubules are in intimate contact with the ER.

(b) If kinesin can be isolated, antibodies could be raised and an immunological assay could be used to localize kinesin to the ER-microtubule junction. Since antibodies against kinesin should be effective at disrupting kinesin-dependent processes, injection of such antibodies into cells could be used to determine whether ER shape and location is adversely affected.

19

Cellular Movement: Motility and Contractility

1. (a) At a sarcomere length of 3.2 μm, the length of the A band is 1.6 μm, and the length of the I band is also 1.6 μm (0.8 μm on either side of the Z line). During contraction of the sarcomere to 2.0 μm, the length of the A band remains fixed at 1.6 μm, while the length of the I band decreases from 1.6 μm to 0.4 μm.

    (b) The H zone corresponds to that portion of the thick filament length that is not overlapped by thin filaments. (It is, in fact, the lack of interdigitated thin filaments that gives the H zone its lighter density and hence its German name.)

    (c) The distance from the Z line to the edge of the H zone represents the length of the thin filament and remains constant during contraction.

2. (a) ATP alone: (0.005 mmol)(1 min/mmol) = 0.005 min, or **0.3 sec.**

    ATP and creatine phosphate: 0.3 sec from ATP plus 1.5 seconds from creatine phosphate (because there is five times as much of it) = **1.8 sec.**

    There are many examples of muscle movement where immediate response rather than sustained contraction is essential. (If you are not convinced that contractions lasting less than 2 seconds are important, try touching a hot stove!)

    (b) (1 mmol ATP/min · g)(1 mmol glucose/36 mmol ATP)(6 mmol O_2/ 1 mmol glucose)(22.4 ml/mmol O_2) = **3.7 ml O_2/min · g.**

(Note that a gram of muscle occupies about 1 cm^3, so the contracting muscle uses more than $3\frac{1}{2}$ times its own volume in oxygen every minute!)

(c) 1% glycogen = 10 mg glycogen/g muscle, so we can write: (10 mg glycogen)(1 mmol glucose/162 mg glycogen)(2 mmol ATP/mmol glucose)(1 min/mmol ATP) = 0.123 min, or **7.4 sec.**

(d) By repeated cycling of ADP through the myokinase reaction, it would, in theory, be possible to convert all the ADP into AMP. In this way, the high-energy anhydride bond of ADP can be used to generate an equimolar amount of ATP, which should therefore sustain contraction for another 0.3 sec (see part a). The question is not realistic, however, insomuch as the cell would be dead if all its ATP were converted to ADP. Furthermore, the myokinase reaction does not "wait" until that stage becomes functional as ATP levels drop and ADP levels climb. Either way, however, it could not sustain contraction for more than an additional 0.3 sec.

3. (a) To determine the amount of calcium ATPase per gram of muscle, note that muscle is 30% protein and that the calcium ATPase represents 1% of the total protein. One gram of muscle therefore contains 0.3 g of protein and 0.003 g (3 mg) of calcium ATPase. With a molecular weight of 115,000, 0.003 g of enzyme represents $0.003/115,000 = 0.026 \times 10^{-6}$ moles, or 0.026 micromoles of enzyme. Since this ATPase can hydrolyze ATP at the rate of 10 micromoles/min per mg, 3 mg of the ATPase will hydrolyze 30 micromoles/min, or 0.5 micromoles/sec. The turnover number of the enzyme is therefore 0.5/0.026 = 19.2 micromoles of ATP hydrolyzed per second per micromole of calcium ATPase, or **19.2 molecules of substrate consumed per second per enzyme molecule.**

(b) Since the total rate of ATP consumption per gram of muscle tissue is 1 mmol/min and the rate of ATP consumption by the calcium ATPase is 30 micromoles/min, the fraction of the total energy expenditure that is used to transport calcium in a contracting skeletal muscle is **0.03, or 3%.**

4. (a) $\Delta G°'$ = –10.3 + 7.3 = –3.0 kcal/mol = –3000 cal/mol.

$\Delta G°'$ = –RT ln K'_{eq} = –3000 cal/mol

ln K'_{eq} = –3000/(–1.987)(273 + 25) = 3000/592 = 5.07

K'_{eq} = $e^{5.07}$ = **158.5.**

(b) $K'_{eq} = \dfrac{[\text{creatine}][\text{ATP}]}{[\text{creatine phosphate}][\text{ADP}]} = 158.5$ or

$\dfrac{[\text{ATP}]}{[\text{ADP}]} = 158.5 \dfrac{[\text{creatine phosphate}]}{[\text{creatine}]}.$

Relative Amounts		Cr-P	ATP	Relative Amounts		ATP
Cr-P	Cr	Cr	ADP	ATP	ADP	ATP + ADP
0.9	0.1	9	1427	1427	1	0.999
0.5	0.5	1	158.5	158.5	1	0.994
0.1	0.9	0.11	17.4	17.4	1	0.945
0.01	0.99	0.01	1.6	1.6	1	0.615

(c) As contraction proceeds, creatine phosphate levels fall very substantially before any perceptible decrease in ATP levels occurs. (For example, when only 10% of creatine is in the phosphorylated form, more than 94% of the ADP still is.) With ATP levels unaffected by consumption of high-energy phosphate during contraction, the conclusion seemed logical that creatine phosphate rather than ATP was the immediate energy source.

(d) With the creatine kinase reaction blocked, creatine phosphate levels would remain high during contraction, and the ATP levels would fall rapidly and precipitously.

(e) This is just what would be expected if ATP were the immediate source of energy in contraction and creatine phosphate were a reserve source used for rephosphorylation of ADP.

5. (a) Rigor results from a failure to break the cross-bridges that link thick filaments to thin filaments in the contraction cycle. In the living cell, detachment occurs upon binding of the next molecule of ATP. After death, however, cellular ATP levels are depleted and cannot be replaced. This has two consequences: (1) ATP is not available to cause cross-bridge detachment, and (2) in the absence of ATP, calcium cannot be pumped into the sarcoplasmic reticulum. Calcium therefore accumulates in the cytoplasm and promotes attachment of cross-bridges to actin. The net result is an accumulation of cross-bridges that lock the muscle filaments together and give the corpse its characteristic stiffness.

    (b) While running; ATP levels would presumably be lower and cross-bridge formation would probably occur more quickly upon death.

    (c) Addition of ATP will have a relaxing effect, because it will allow cross-bridges to break and filaments to detach.

6. (a) The cross-bridges will be dissociated from the thin filaments, but will not be "recocked" to the configuration normally seen in resting muscle, because AMPPCP cannot be hydrolyzed to ADP and P_i. Your diagram should look like that after step 3 in the contraction cycle of Figure 19-14 except that AMPPCP is bound to the myosin head instead of ATP.

    (b) Yes, because as a structural analogue, AMPPCP is presumably a competitive inhibitor of the ATPase and should be displaced from the binding site by a high enough concentration of ATP.

    (c) All other ATP-dependent processes are likely to be inhibited, including uptake of calcium ions by the sarcoplasmic reticulum.

7. (a) As a physiological phenomenon, tetanus results from the arrival of successive impulses at the muscle cell membrane at such a rapid rate that calcium ions cannot be pumped back into the SR between pulses. These ions thus remain in the cytoplasm, causing

continuous contraction and leading to a "cramping" of the muscle in the fully contracted form that persists as long as the rapid nerve impulses do.

(b) The exotoxin produced by *C. tetani* could be a powerful stimulant of the central nervous system, such that nerve impulses are sent to the muscle at a rapid, unremitting rate. Or it could act directly on the sarcoplasmic reticulum, causing release of (or failure to accumulate) calcium. (The former explanation is in fact correct.)

(c) The bacterium normally gains entry into tissue through wounds, and puncture wounds are most likely to produce the anaerobic conditions needed by the organism.

8. (a) A, H   (c) A, H, I, R   (e) A, H, I   (g) A

    (b) R   (d) I   (f) H

9. (a) Colchicine binds to unpolymerized tubulin, preventing it from polymerizing into microtubules in the flagella.

    (b) The pool of unpolymerized tubulin in the cytoplasm is too small to produce enough microtubule material for flagella of normal length. Synthesis of tubulin is required to enlarge the pool.

    (c) This phenomenon is caused by depolymerization and repolymerization of tubulin in equilibrium with the cytoplasmic tubulin pool.

10. (a) Protein A is most likely the structural subunit of the microtubule, since it has no ATPase activity and is of the right molecular weight for tubulin. Protein C is probably dynein, the protein of the arms. It has ATPase activity, associates with "de-armed" doublets, and restores the arms. Protein B is probably a subunit of dynein.

(b) Since de-armed doublets are apparently available, it would be interesting to see if they consist of protein A only and whether they can be reconstituted from a solution of protein A. It would also be relevant to attempt to dissociate protein C and to determine whether protein B is among the polypeptides of which protein C is composed.

11. (a) The nonmotility of the sperm cells (and hence the sterility of the individual) is due to one or more structural defects in the sperm tail, causing it to be nonfunctional. (A variety of such defects have in fact been reported, including lack of both dynein arms, lack of only the outer or inner arm, lack of the radial spoke heads, or lack of one or both microtubules of the central pair.)

(b) The same structural defects in the sperm tail are also likely to affect the cilia responsible for sweeping mucus and foreign matter out of the lungs and sinuses.

(c) Since the sperm are nonmotile, you should not expect any offspring at all!

20

Electrical Signals: Nerve Cell Function

1. (a) S   (c) S   (e) N   (g) N

    (b) A   (d) N   (f) S   (h) S

2. (a) E_{Cl} will be negative, because only a negative charge on the inner surface of the membrane would counteract the tendency of the concentration gradient to drive chloride ions inward.

    (b) $E_{Cl} = (RT/zF) \ln \dfrac{[Cl^-]_{out}}{[Cl^-]_{in}} = -58 \log_{10} \dfrac{560}{50} = -60 \text{ mV}.$

     (Note that the RT/zF value is negative because z = –1 for an anion.)

    (c) If the internal chloride ion concentration is 150 mM instead of 50 mM, the value of E_{Cl} becomes less negative (–33 mV instead of –60 mV).

    (d) The chloride ion concentration is probably so variable because the positive charge of sodium and potassium ions can be balanced in part by other anions, notably proteins. To the extent that the protein content of the axon varies, so might the internal chloride concentration.

3. (a) Using the known intracellular potassium ion concentrations of Table 20-1, equation 20-1 can be rearranged to express E_K in terms of temperature:

$$E_K = \frac{(2.303)(1.987)T}{23,062} \log_{10} 0.05 = -0.0002584T$$

At 18°C (= 291 K):

$$E_K = -0.0002584(291) = -0.0752 \text{ V} = -75.2 \text{ mV.}$$

At 25°C (= 298 K):

$$E_K = -0.0002584(298) = -0.0770 \text{ V} = -77.0 \text{ mV.}$$

For most purposes, the difference of 1.8 mV would probably be of modest significance only.

(b) At 37°C (= 310 K):

$$E_K = -0.0002584(310) = -0.0800 \text{ V} = -80.0 \text{ mV.}$$

The difference of almost 5 mV would probably be significant in most experiments.

A temperature of 37°C would be more meaningful for experiments with mammalian neurons, since this closely approximates the body temperature of many warm-blooded animals.

4. (a) These are the only three ions to which the plasma membrane of the nerve cell is sufficiently permeable to warrant their inclusion in the equation.

(b) A more general expression for monovalent ions would be

$$V_m = \frac{RT}{F} \ln \frac{\sum P_{cation} [cation]_{out} + \sum P_{anion} [anion]_{in}}{\sum P_{cation} [cation]_{in} + \sum P_{anion} [anion]_{out}}$$

(c) No; calcium accounts for the main ion flux across the membrane of the SR, so the expression would also need a term for divalent cations.

(d) With the relative permeability for sodium ions at 0.04, the value for V_m is –60 mV, calculated according to equation 20-6. If the relative permeability for sodium ions were 1.0 instead, the value for V_m would be about –9 mV.

(e) No, because V_m is proportional to the logarithm of the expression that contains the term for sodium permeability.

5. (a) Of the sodium channels in the membrane, at least some will open in response to a stimulus that depolarizes the membrane by about 20 mV.

(b) Depolarizations of less than 20 mV are not sufficient to start the positive feedback cycle that leads to an action potential, and they therefore elicit no response.

(c) Intensity of stimulus is detected as the frequency with which individual neurons respond and/or a difference in the number of separate neurons that respond.

6. (a) B, C   (d) A, F   (g) C

(b) C, D, E   (e) None   (h) B

(c) C   (f) A, B, C, D, E, F   (i) E

7. (a) (5 pA)(1 × 10⁻¹² ampere/pA)(6.2 × 10¹⁸ charges/ampere)(0.005 sec) = 155,000 charges. Thus, 155,000 ions pass through a single channel during the 5 milliseconds that it is open.

(b) No. Even though 155,000 ions pass through a single open channel, this is insufficient current to cause any perceptible change in the resting potential of a postsynaptic membrane. (An actual synaptic transmission event involves the simultaneous fusion of several hundred synaptic vesicles with the presynaptic membrane. Each vesicle releases several thousand neurotransmitter molecules, and each of these, in turn, caused thousands of receptor channels in the postsynaptic membrane to open for a few milliseconds. Thus, several hundred thousand channels open at once, generating a current that is sufficient to drive the resting potential of the postsynaptic membrane to its threshold, leading to depolarization of the membrane and an ensuing action potential.)

8. (a) $E_K = \dfrac{(2.303)(1.987)(273 + 37)}{(+1)(23,062)} \log_{10} \dfrac{4.6}{150}$

$= +0.06157 \log_{10} (0.0306)$

$= -0.0932 \text{ V} = -93.2 \text{ mV.}$

$E_{Na} = +0.06157 \log_{10} \dfrac{145}{10} = +0.0715 \text{ V} = +71.5 \text{ mV.}$

$E_{Ca} = +0.06157/2 \log_{10} \dfrac{6}{0.001}$

$= +0.0308 \log_{10} (6000)$

$= +0.116 \text{ V} = +116 \text{ mV.}$

(b) The equilibrium potential for potassium is substantially more negative (–93.2 mV instead of –75 mV) because of a greater concentration gradient.

(c) Either Na^+ or Ca^{2+}, or both; both would move *inward.*

(d) You could remove either sodium or calcium ions from the surrounding medium and observe the effect this has on the action potential. Radioactive isotopes could also be used to follow specific ions.

(e) Potassium is driven outward both by its concentration gradient across the membrane and by the temporarily positive membrane potential. At A, the voltage-dependent potassium gates are not yet open.

(f) This should decrease the concentration gradient for potassium and therefore decrease the rate of potassium ion movement out of the cell. The resting potential for the cell should be less negative.

9. (a) Competes with acetylcholine for binding to the acetylcholine receptor, thereby blocking depolarization of the postsynaptic membrane; leaves membrane fully polarized.

    (b) Forms a covalent complex with the active site of acetylcholinesterase, thereby preventing hydrolysis of acetylcholine; leaves membrane depolarized.

    (c) Mimics acetylcholine in binding to the acetylcholine receptor and causing depolarization of the postsynaptic membrane, but is hydrolyzed much more slowly; leaves membrane depolarized.

    (d) Inhibits acetylcholinesterase by forming a stable carbamoyl-enzyme complex at the active site, thereby preventing the hydrolysis of acetylcholine; leaves membrane depolarized.

21
Chemical Signals: Hormones and Receptors

1. (a) Calmodulin

   (b) pituitary

   (c) ligand

   (d) inositol trisphosphate (InsP$_3$) and diacylglycerol (DAG)

   (e) adenylate cyclase; phosphodiesterase

2. (a) F; receptors are found on most if not all cells.

   (b) F; receptors have not yet been definitively identified for plant cells, but plant growth and development is known to be regulated by a number of hormones, so plant cells presumably have receptors to detect these hormones.

   (c) F; hormones secreted by endocrine glands are released directly into the circulating blood.

   (d) T; hormones such as epinephrine affect smooth muscle.

   (e) F; hormones are synthesized either from amino acids or from cholesterol.

   (f) F; steroid hormones bind directly to their intracellular receptors rather than acting through second messenger systems.

   (g) F; calmodulin is a peptide but not a hormone.

(h)   F; inositol trisphosphate is produced by lipid hydrolysis but is not itself a lipid. (DAG, the other hydrolysis product, is a lipid, however.)

3.   (a)   **Nervous system**

Advantage: Communication is fast and direct.

Disadvantage: Direct communication is possible only between connected cells.

Endocrine system

Advantage: Communication is possible with all cells in body.

Disadvantage: Communication is slow and not readily down-regulated.

Paracrine system

Advantage: Provides means of local communication and regulation.

Disadvantage: Slow, relatively nonspecific, not readily down-regulated. (Allergic reactions, for instance, are the result of an inappropriate paracrine response caused by histamine release from mast cells.)

(b)   When all three systems act in concert, the advantages of each system tend to balance the disadvantages of the others, providing a functional means of regulating physiological responses.

4.   The following is a functional feedback loop that allows pancreatic islet cells to regulate blood glucose:

Step 1:   When blood glucose concentrations are increased (after eating a meal rich in carbohydrates, for example), more glucose is available in the circulating blood to bind to glucose receptors on the plasma membrane of pancreatic islet cells.

Step 2:   Binding of glucose to its receptors causes the plasma membrane to depolarize more frequently.

Step 3:   Membrane depolarization causes vesicles containing insulin to fuse with the plasma membrane and release their contents exocytotically into the circulating blood.

Step 4:   The insulin travels through the bloodstream to target cells, such as those of muscle tissue, with insulin receptors on the plasma membrane.

Step 5:   Binding of insulin to its receptors activates the inward transport of glucose.

Step 6:   Glucose moves into cells of the target tissues, thereby lowering the concentration of glucose in the blood.

Step 7:   As the glucose concentration in the circulating blood decreases, glucose dissociates from membrane receptors on islet cell membranes, thereby causing a decrease in depolarization rates and reduced release of insulin.

5.   The following is a pathway that mediates the response to fright by linking the initial detection of danger by the brain to an increased heart rate:

Step 1:   The brain receives sensory information (from the ears, eyes, nose, or other sensory systems) that indicates danger.

Step 2:   A nervous signal is sent down the spinal cord and out along a peripheral nerve which connects to and regulates the activity of the adrenal gland.

Step 3:   The nerve signal causes cells in the adrenal medulla to release epinephrine into the circulating blood by exocytosis.

Step 4:   The epinephrine travels via the bloodstream to the heart, where it binds to epinephrine-specific membrane receptors on target cells.

Step 5:   The receptors activate a pathway that causes an increase in the rate of beating by individual heart cells and hence by the heart itself.

6. Given that transmembrane channels transport water across the plasma membrane of kidney tubule cells, antidiuretic hormone (ADH) must somehow affect these channels. One possibility is that ADH causes increased numbers of channels to be inserted into the membrane, thereby permitting more water to leave the tubule and be transported back into the blood. As a result, less water would pass through the tubules into the urine, which is just the effect required if ADH is to reduce loss of water from the body.

7. Secretory cells of the respiratory tract secrete mucus when adenylate cyclase is activated by a G protein with a GTP bound to it. Normally, this system is down-regulated when the GTP is hydrolyzed to GDP. This down-regulation is inhibited by pertussis toxin, which alters the G protein so that the bound GTP cannot be hydrolyzed. The system therefore remains activated and copious quantities of mucus are secreted into the respiratory tract, giving rise to the symptoms of whooping cough. (Pertussis toxin is actually an enzyme that cleaves NAD^+ and transfers the ADP-ribose portion of the coenzyme to the α subunit of the G protein. The GTPase activity of the G protein is thereby irreversibly inhibited, leaving the adenylate cyclase permanently activated.)

8. Most cells are able to maintain a very low intracellular Ca^{2+} concentration (in the range of 10^{-8} to 10^{-7} M) by using a calcium ATPase to move calcium ions outward. Because the intracellular calcium concentration is so low, it can be changed dramatically and rapidly by admitting very small amounts of calcium and the concentration change can very effectively regulate specific cellular functions. For example, a ten-fold concentration in calcium concentration in a muscle cell can be achieved within milliseconds by releasing micromolar amounts of calcium ions from the sarcoplasmic reticulum into the cytoplasm. To achieve a ten-fold change in the concentration of a more abundant ion such as potassium would require the release of molar amounts of potassium ions within milliseconds, which is impossible. Another candidate for a regulatory role is hydrogen ion, which is present in

most cells at about the same concentration as calcium (10^{-7} M at pH 7.0, for example). Some cells actually use changes in the hydrogen ion concentration to regulate certain activities, but this is not a generally useful regulatory system because hydrogen ions bind so universally to acidic and basic groups on all proteins, thereby affecting many physiological processes indiscriminately. In contrast, proteins such as calmodulin have highly specific binding sites for calcium ions and are therefore very sensitive to changes in calcium concentration that have little or no effect on other proteins.

9.   The binding of epinephrine to beta adrenergic receptors causes a stimulation of heart function, both in terms of heart rate and with respect to the amount of work done in pumping blood. The effect appears to be mediated by cyclic AMP. When a beta blocker (such as propanolol) is given to patients with hypertension, the cellular response caused by the binding of epinephrine to beta receptors is partially inhibited. Heart function is gradually restored over a period of time, with a corresponding decrease in blood pressure.

10. The first step is to choose the simplest system that will allow you to do the required experiments. You might, for example, choose yeast cells first, since they are easy to grow and are genetically well characterized. For a photosynthetic cell, an alga might be best. A higher plant would be used only if the question cannot be adequately addressed in a simpler system. Next, you would decide on a function that can be experimentally manipulated, presumably some chemical or environmental signal to which the cell clearly responds. An initial experimental approach might be to look for proteins that behave physically like G proteins. You might, for instance, isolate plasma membranes and test for sites to which GTP and GDP can bind. Next would be a functional test: Do the GDP and GTP cycle on and off the site in response to the desired signal? A positive result here would make the project quite interesting. You might then subject isolated plasma membranes to an isolation procedure known to yield purified G proteins from animal cells and see if a protein can be isolated that has

similar properties to known G proteins. Assuming a positive result here, you might go on to isolate the gene(s) that encode the isolated protein, using the methods described in Appendix B. If the genes turn out to have strong homologies to known genes for G proteins in animal tissues, you would have achieved a significant result that would be likely to appear in the next edition of this textbook.

22
Cellular Aspects of Embryonic Development

1. (a) Germ cells give rise to gametes, and thus to the next generation. Somatic cells form all the various tissues of a multicellular organism, including most of the gonads.

   (b) The two are closely coupled. The acrosomal reaction: sperm contacts egg, release of acrosomal vesicle contents, enzymatic digestion of the jelly coat, actin polymerization to form the acrosomal process, reaction of bindin on the acrosomal process with receptor on egg membrane, fusion of egg and sperm membranes. The cortical reaction: membrane depolarization and change in membrane potential (triggered by membrane fusion), calcium wave, release of cortical vesicles, enzymatic thickening and hardening of the perivitelline space.

   (c) The blastula, a mass of cells, is the embryonic stage at the end of cleavage. The cells of the blastula surround the blastocoel cavity.

   (d) Whereas the mere presence of a cytoplasmic determinant specifies a single cellular fate, a morphogen gradient specifies several cellular fates, each determined by the particular concentration of the morphogen.

   (e) Differentiation is the manifestation of a cellular fate, such as the expression of myosin and actin, filament assembly, and cell fusion in muscle cells. Determination precedes differentiation and means that a cell has irreversibly entered a specific developmental pathway, although it does not yet show any physical signs of differentiation. Myoblasts, which differentiate into muscle cells, are a determined cell type.

(f)   The segmentation gene products divide the *Drosophila* embryo into segments, which are then assigned specific and distinct developmental fates by the products of the segment-identity genes. Without the latter, the embryo would develop with 14 identical segments.

(g)   Homeo proteins are nuclear proteins that function by binding to, and thus regulating, the DNA of specific genes. Growth factors are secreted extracellular proteins that regulate cell growth of particular cell types.

(h)   A mutation is cell-autonomous if it affects only the cell(s) where the mutant gene product is normally made. A mutation is cell-non-autonomous if it affects cell(s) other than the ones that normally make the gene product. Thus a mutation in a homeo-protein gene is likely to be cell-autonomous, while a mutation in a growth factor gene is likely to be cell-nonautonomous.

2.   (a)   Cassette switching at the *MAT* locus in the yeast mating type switch; immunoglobulin gene rearrangement (see Chapter 24).

(b)   18S and 28S ribosomal RNA genes in oocyte development.

(c)   Action of acrosomal and cortical vesicles during fertilization.

(d)   Masking of maternal mRNAs and inactivation of ribosomes in oocytes.

(e)   Flagellar motion of the sperm tail; changes in cytoskeleton that mediate changes in cell shape; mitotic spindle.

(f)   Extension of the acrosomal process; changes in cytoskeleton that mediate changes in cell shape.

(g)   Regulation of the cell cycle.

(h)   Invagination of the archenteron; neural crest and primordial germ cell migrations.

(i)   Homeo and zinc finger proteins made by *Drosophila* pattern formation genes.

(j)   Yeast cell cycle arrest upon binding of peptide mating factors; cAMP and DIF action in *Dictyostelium*; any inductive interaction; inhibition of meiosis in *Caenorhabditis* germ cells by the distal tip cell.

3.   (a) D, K, P   (d) M   (g) B, D, J, M   (j) J

     (b) B, J, M   (e) D   (h) B, J, M

     (c) K, P   (f) B, M   (i) N

4.   Correct order: a, e, g, c, f, b, d, h.

5.   (a) Stage 1

     (b) Stages 4 and 8

     (c) Stage 6

     (d) Neural tube formation

     (e) Correct order: 3, 7, 5, 1, 8, 4, 6, 2.

6.   (a) Inject the dye into single blastomeres and record what tissues of the developing embryo and tadpole contain the dye.

     (b) The small chemical dye will diffuse through the gap junctions, and some cells that are *not* descendants of the injected cell will contain the dye. Horseradish peroxidase is a large protein that cannot diffuse through gap junctions.

(c) Polarity is not yet established at the 8-cell stage of mouse (and other mammalian and avian) embryos. Therefore individual cells cannot be distinguished for injection. Even if they could be, the results would not be reproducible.

7. (a) Yes; transcription from *HMRa* and *HMLα* is silenced by the action of the SIR gene products, which bind to specific sequences in the DNA flanking the *a* and *α* cassettes.

(b) No; the genes are amplified to meet transcriptional needs during oogenesis. Extra gene copies are then simply lost during development.

(c) Yes; the extra copies of the 5S gene needed during oogenesis are a permanent part of the genome. Most 5S gene transcription must be actively shut down in somatic cells.

(d) No; the oocyte is a single cell, thus the mRNA must be localized by mechanisms other than differential transcription (see Problem 9).

(e) Yes; the pair-rule genes are regulated by the gap and polarity genes, many of which encode transcription factors. Thus, the striped pattern is the result of transcriptional regulation, but it is unclear whether this is positive or negative regulation.

8. Yes; if MPF/cdc2 is stable and is activated by being phosphorylated, then there must be phosphatases around to remove the phosphoryl groups, and thus inactivate MPF/cdc2, preparing it for the next cell cycle. Similarly, since MPF/cdc2 regulates other proteins by phosphorylating them, there must be phosphatases that counter the action of MPF/cdc2, and thus allow the cell cycle to continue.

9. (a) The globin mRNA is a control. It shows that injected mRNAs do not automatically accumulate in the vegetal hemisphere.

(b) The Veg1 mRNA could be either transported to the vegetal hemisphere or degraded in the animal hemisphere.

(c) Degradation in the animal hemisphere.

(d) Association with ribosomes and/or translation is not necessary for localization.

(e) Cytoskeleton-associated proteins must recognize and bind the Veg1 mRNA. Perhaps Veg1 mRNA is transported to the vegetal hemisphere along filaments of the cytoskeleton. To test, co-inject Veg1 mRNA with colchicine or cytochalasin. If the hypothesis is correct, one of these two drugs should stop transport and tell you which cytoskeletal components are involved.

10. (a) Absolutely not! All we know is that XGP is localized near the point of sperm entry. We have no evidence that XGP mediates polarity.

(b) Inject purified or in vitro-synthesized XGP into other sites in the fertilized egg. If XGP establishes polarity, this should cause duplication of structures.

(c) (i) XGP could be an enzyme involved in the biosynthesis of a chemical morphogen (such as DIF, cAMP or retinoic acid).

(ii) XGP could be a transcription factor. As cleavage proceeds, cells having a high XGP concentration will transcribe genes for the dorsal cell fate, while cells with a low XGP concentration will transcribe genes for more ventral cell fates.

(iii) XGP could be a cell surface receptor that is incorporated into the membranes of dorsal cells as cleavage proceeds, making those cells more responsive to some ubiquitous chemical messenger. More ventral cells have progressively less receptor and are thus progressively more ventral.

(iv) XGP could be the ligand for a ubiquitously distributed receptor. (Note that this is the converse of #3.)

(d) Differential translation: sperm entry activates translation of nearby XGP mRNA. Veg1 localization is carried out pretranslationally.

11. (a) No; N-CAM mRNA is not found until after induction has begun, thus N-CAM probably does not participate in the response to inducer.

(b) Treat isolated neural plate ectodermal cells with the putative inducer, and look for N-CAM mRNA. Anything that causes N-CAM mRNA to be expressed is a possible inducer.

(c) In neural induction, the mesoderm is the inducing tissue, thus the polypeptide should be made in mesodermal cells.

(d) (i) We must show that the polypeptide is produced in vivo by the mesodermal cells of the neural plate.

(ii) To meet the criteria of induction, we must show that it can cause neural tube formation in equivalent neural ectoderm cells.

12. (a) In the absence of mesenchyme (the inducing tissue) the salivary epithelium (the induced tissue) adopts a different cellular fate.

(b) Yes; because branching only occurs after a new layer of GAGs has been made. GAGs could either be the receptor for the induction signal or could trigger release of the induction signal from the mesenchymal cells.

(c) The cells continue to express GAGs, which is a manifestation of their inherited differentiated state as salivary epithelial cells.

(d) During branching, cell shape changes significantly. Thus, induction must cause changes in the cytoskeleton. Since salivary glands are secretory tissue, induction also must activate secretory pathways.

13. (a) The mutations would otherwise kill the cells.

(b) (i)   A promoter mutation could prevent transcription.

(ii) A missense mutation could change an amino acid in the "tail" of the protein and prevent assembly of the mutant myosin into filaments.

(iii) A missense mutation could change an amino acid in the "head" of the protein, thus preventing the head from binding actin filaments during the contraction cycle (or it could disrupt ATPase).

(iv) A nonsense mutation could make a myosin heavy chain with a shortened tail. (The tail is encoded by the 3' end of the gene.)

(c) No; because normal and mutant actin monomers are present in a 1:1 ratio. Thus every growing filament will soon incorporate a mutant actin, and cease growing. If, however, the mutant actin cannot polymerize, then the normal actin will form normal microfilaments.

(d) P   EMS-treated w/Y males × +/+ females

F$_1$ individual w*/+ females × w/Y males

F$_2$ classes: w*/Y males

+/Y males

w*/w females

+/w females

If the w*/Y males are missing, then there is a zygotic lethal mutation. The mutant chromosome can be rescued from the w*/w females. The mutagenesis takes two generations instead of three.

(e) P   Allow mutagen-treated unc hermaphrodites to self-fertilize.

F1   Allow individual */+ heterozygotes to self-fertilize.

F2   classes: unc*/unc*, unc */+, and +/+.

If the unc*/unc* class is missing, there is a zygotic lethal mutation on the unc chromosome, and the mutant chromosome is present in the *unc* */+* animals. With self-mating, all you have to do is culture individual F1 worms, and examine the F2 generation under the microscope. No matings to set up!

14. (a) Since it is expressed in dorsal tissues, and since in its absence, dorsal ectodermal cells develop as ventral ectodermal cells, *dpp* must be involved in specifying or expressing the dorsal ectodermal cell fate.

(b) The *dl* gene, because it is maternally supplied, whereas *dpp* is made in the zygote.

(c) The *dl* transcriptional regulator could either inhibit *dpp* mRNA expression in the ventral cells or it could activate *dpp* mRNA expression in dorsal cells.

15. (a) Differential splicing; use of different promoters; use of different poly-A sites.

(b) The 700 base intron that interrupts the open reading frame of the male mRNA is spliced out in the female mRNA, creating a 3.7-kb mRNA.

(c) When the male mRNA is translated, only a truncated and nonfunctional protein is made. When the fully spliced female mRNA is translated, a complete, functional protein is made.

(d) Males are unaffected because the gene product performs no function in males.

(e) The Sex lethal protein mediates the splicing of similar introns from these mRNAs in females. In males, where no functional Sex lethal protein is made, the introns are not spliced out and the proteins perform no (or different) functions. The other proteins are involved in carrying out female sexual differentiation.

23
Cellular Aspects of Cancer

1. (a) A benign tumor is noninvasive and nonmetastatic; a malignant tumor is invasive and metastatic.

   (b) Normal 3T3 fibroblasts grow to confluence in culture, then growth ceases; SSV-transformed 3T3 cells continue growing after confluence.

   (c) Viral oncogenes promote the development of the neoplastic phenotype when introduced into suitable recipient cells; proto-oncogenes that are under normal control fail to induce neoplasia. In addition, viral oncogenes lack introns whereas proto-oncogenes usually have introns.

   (d) Mitogens that bind to the EGF receptor will inhibit the binding of radiolabeled authentic EGF to cells; mitogens that do not bind to the EGF receptor will not affect the binding of radiolabeled EGF.

   (e) The normal granulocyte will have a normal karyotype; a chronic myelogenous leukocyte will have the Philadelphia chromosome.

2. The experimental analysis consists of two types of experiments:

   (a) One must show that there is a consistent association of the virus with the tumor: All tumors must be shown to contain the virus whereas normal, nondiseased animals lack the virus. The presence of virus is usually shown:

      (i) Serologically: Tumor-bearing animals contain antibodies to viral proteins.

(ii)   By nucleic acid hybridization: Tumor cells contain viral nucleic acid that is demonstrated by molecular hybridization with appropriate probes.

(b)   One must satisfy Koch's postulate. Virus particles isolated from a tumor should induce the same tumor when inoculated into a disease-free host. If it proves possible to isolate the same virus from the induced tumor and to show that it causes the same tumor type when inoculated into a second set of disease-free hosts, Koch's postulate is said to have been demonstrated.

3.   Evidence for carcinogenicity:

Category 1.   Epidemiologic studies of human populations have shown that individuals who suffer chronic exposure to a particular carcinogen have a significantly higher incidence of cancers caused by that carcinogen than do a matched control population who are not exposed to the carcinogen.

Category 2.   Experimental studies of experimental animals have shown that animals that have been deliberately exposed to the carcinogen have a higher incidence of particular cancers than control animals that were not exposed.

Evidence for mutagenicity:

Mutagens increase the rate of back mutation of auxotrophic mutant strains of bacteria (i.e., the bacteria will grow only if the medium is supplemented to contain the metabolite that they are incapable of synthesizing, due to the mutation of one of the enzymes of the biosynthetic pathway of that metabolite) to convert them to prototrophy (i.e., the bacteria can grow on medium lacking the metabolite whose biosynthesis was defective in the mutant individuals).

4.   (a)   *Angiogenesis:* The tumor becomes invaded by blood vessels.

(b)   *Invasion:* Tumor tissue invades across the vascular endothelium to establish direct contact with the blood.

(c) *Release:* Clumps of tumor cells break free of the parent tumor mass and enter the blood.

(d) *Transport:* The circulating clumps of tumor tissue are transported by the blood.

(e) *Lodgement:* The circulating clumps of tumor cells arrest in a capillary bed in an organ distant from the location of the primary tumor.

(f) *Invasion:* Cells of the secondary tumor invade across the endothelium of the capillary into the stroma of the host organ.

(g) *Growth:* The secondary tumor enlarges by growth in the stroma of the new host organ.

5. (a) *Cell fusion:* Hybrid cells derived by the fusion of non-neoplastic and neoplastic parents are non-neoplastic (i.e., the cells fail to form tumors when introduced into an appropriate host animal). When particular chromosomes of the normal chromosomal set are lost, the hybrid cells revert to neoplasia. This reversion can be reversed (i.e., the cells return to the non-neoplastic condition) when that particular normal chromosome is introduced back into the cell.

(b) Certain human tumors have been associated with the establishment of loss-of-function mutations of particular genes. When a functional copy of that gene is introduced into the tumor cells, they revert to the non-neoplastic phenotype.

6. (a) A variety of chemicals have been shown to be carcinogenic, both by retrospective epidemiological studies on exposed populations of humans, and by prospective experimental studies of tumor incidence in experimental animals deliberately exposed to the carcinogen.

(b) All human cancers show profound differences in frequency in different ethnic and national groups. These differences in frequency are due mainly to nongenetic differences between these populations because emigrant populations lose the frequency profile of the homeland, and acquire the frequency profile of the host country much faster than can be accounted for by the rates of gene mixing by intermarriage. Thus, the variation in the ethnic and national patterns of tumors is largely ascribable to differences in exposure to environmental factors.

7. (a) Solid tumors are dependent on the presence of a vasculature for growth. When it is possible to inhibit selectively the development and maintenance of the vasculature of the tumor, it regresses and becomes significantly less threatening.

    (b) Since the exposure to environmental carcinogens plays an essential role in the initiation of many human tumors, it should be possible to reduce the incidence of tumors if it were possible to reduce the extent of contact between people and chemical carcinogens. To accomplish this, we first need to know which chemicals are carcinogenic and which are not.

8. (a) Gene mutation and chromosomal rearrangement provide the genetic background for the establishment of novel characteristics required for increased malignancy.

    (b) The ability to attract vascular capillaries from surrounding host tissues is necessary for the growth of solid tumors. In the absence of a vasculature, solid tumors remain small and benign.

    (c) A reduction in the requirement for exogenous supplies of the trophic hormones and mitogens necessary to stimulate proliferation of cells of the normal tissue renders the tumor independent of the host's growth regulatory machinery.

(d) Reduced susceptibility to the immune defense processes of the host that recognize and destroy tumor cells allows a tumor to grow and metastasize without hindrance.

(e) Increased motility, decreased cell adhesion, and increased release of proteolytic enzymes contribute to increased potential for invasion and metastasis.

24
Cellular Aspects of the Immune Response

1. (a) T

   (b) T

   (c) F; removal of the bursa of Fabricius will primarily impair the humoral immune response.

   (d) T

   (e) F; CDC4 T cells are class II-restricted.

2. (a) Since irradiation of the animal destroyed both its white blood cells and its immune response, the immune response is probably dependent in some way on the presence of white blood cells.

   (b) These experiments were done to determine which type of white blood cells could restore the immune response in the irradiated animals, thereby establishing which type accounts for the response in normal animals. The results indicate that lymphocytes are the specific white blood cells responsible for restoring the immune response.

   (c) Both the humoral and the cellular responses will be restored because both the B and T lymphocytes are white blood cells.

   (d) If the thymus had been removed, both the cellular and the humoral immune responses would have been impaired, but the former more so than the latter, because T lymphocytes are produced in the thymus.

(e) The donor mouse must be of the same inbred line so that the lymphocytes obtained from it will recognize the antigens of the irradiated host as self and not raise an immune response against them.

(f) If lymphocytes from another source had been used, they would have recognized the antigens of the irradiated animal as foreign and would have mounted an immune response against them that would probably have killed the host animal. (This is called *graft-versus-host disease*.)

3. (a) It probably seemed like a logical, straightforward way to explain how antibodies could be raised against an incredible variety of antigens without postulating as many antibody genes as there are distinguishable antigens.

(b) The instructional hypothesis was discredited when it was shown that the three-dimensional structure of a protein molecule depends uniquely on its amino acid sequence and always renatures to the same configuration.

(c) To disprove the instructional hypothesis directly, simply denature an antibody molecule and demonstrate that it always refolds in the same way, regardless of what antigens are present.

4. (a) V_H and V_L (d) C_H

(b) V_H and V_L (e) C_H

(c) C_H

5. HIV infects CD4 T cells by binding to the CD4 glycoprotein. Loss of CD4 T cells results in severe immunodeficiency because CD4 T cells are required for activation of both CD8 T cells and B cells. CD8 T cells mediate cellular immunity via cytotoxicity, whereas B cells mediate humoral immunity via antibody production.

6.   (a) L   (c) H   (e) L   (g) LH

     (b) LH   (d) N   (f) LH

7.   (a) Myeloma DNA can be digested with a restriction enzyme and the restriction fragments tested for their ability to hybridize with a cDNA probe that has been prepared in vitro by reverse transcription of the C region sequence of mRNA molecules specifying the specific myeloma H chains.

     (b) The order of the genes is C_m-C_o-C_q-C_p-C_n. The gene for C_m is closest to the genes for the variable region, since it is the first C_H gene expressed. Both the C_m and C_o genes are deleted in a cell that is expressing C_q (that is, a cell that is making IgQ), so C_o must be between C_m and C_q. Since all other C_H genes are deleted in a cell that is expressing C_n (that is, a cell that is making IgN), C_n must be the most distal gene and C_p is therefore adjacent to it.

8.   (a) Number of V_K genes: $500 \times 5 = 2500$.

     Number of V_H genes: $400 \times 20 \times 4 = 32{,}000$.

     Number of antibody molecules: $2500 \times 32{,}000 = 8 \times 10^7$.

     (b) The C_K and C_H genes are not involved in the antigen-binding site and therefore do not contribute to the diversity of antigens that can be recognized.

     (c) In each of the three cases (V and J joining to form the V_K gene and both V-D and D-J joining to form the V_H gene), all three possible reading frames can be generated, so the additional diversity is $3 \times 3 \times 3 = 27$-fold, for a total of $(27)(8 \times 10^7) \approx 2 \times 10^9$ different kinds of antibodies.

     (d) Further diversity is possible as a result of somatic mutations in any of the variable-gene segments—mutations that occur in any of the V, J, and D segments of the lymphocyte genome and are therefore not passed on to the next generation, but can be expressed in antibody formation.

(e) Antibodies and T cell antigen receptors use all of the same mechanisms for the generation of diversity except for somatic hypermutation, which is found in antibodies but not in T cell antigen receptors.

9. (a) The T_c cells are active against the virus-infected fibroblasts because this is the same virus that was used to activate the T_c cells.

(b) The T_c cells are not active against the virus-infected fibroblasts because this is a different virus than that used to activate the T_c cells.

(c) The ineffectiveness of the T_c cells in this case must mean that the T_c cells recognize not just the virus, but some substance present on fibroblasts of strain 1 but not on fibroblasts of strain 2.

(d) The dependence of cytotoxic effect on genetic identity at the MHC loci and its insensitivity to genetic differences at all other loci suggest that class I MHC glycoproteins are somehow required for T_c cells to recognize viral antigens.

TRANSPARENCY MASTERS

TABLE OF CONTENTS FOR TRANSPARENCY MASTERS

TABLE OF CONTENTS (continued)

The *World of the Micrometer*

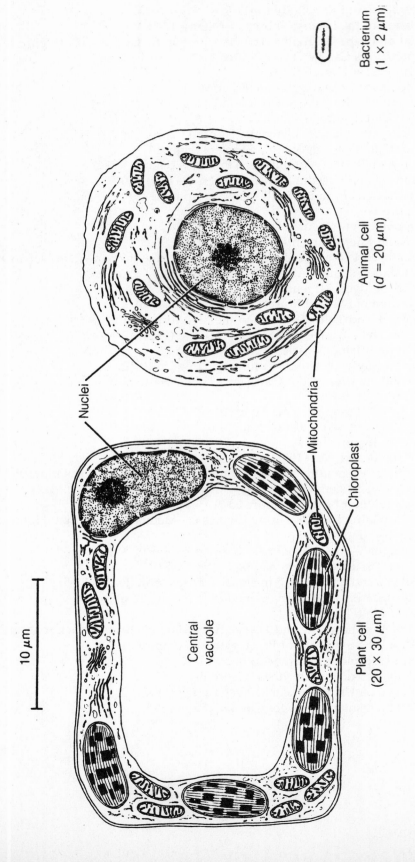

10 μm

Nuclei

Central
vacuole

Mitochondria

Chloroplast

Plant cell
(20 × 30 μm)

Animal cell
(d = 20 μm)

Bacterium
(1 × 2 μm)

Copyright © The Benjamin/Cummings Publishing Company, Inc.

The World of the Nanometer

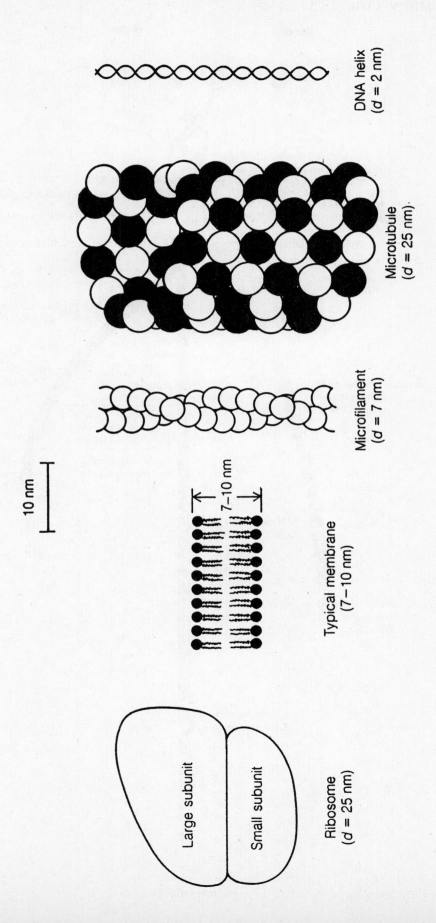

DNA helix
(d = 2 nm)

Microtubule
(d = 25 nm)

Microfilament
(d = 7 nm)

10 nm

7–10 nm

Typical membrane
(7–10 nm)

Large subunit

Small subunit

Ribosome
(d = 25 nm)

Copyright © The Benjamin/Cummings Publishing Company, Inc.

The Cell Biology Time Line

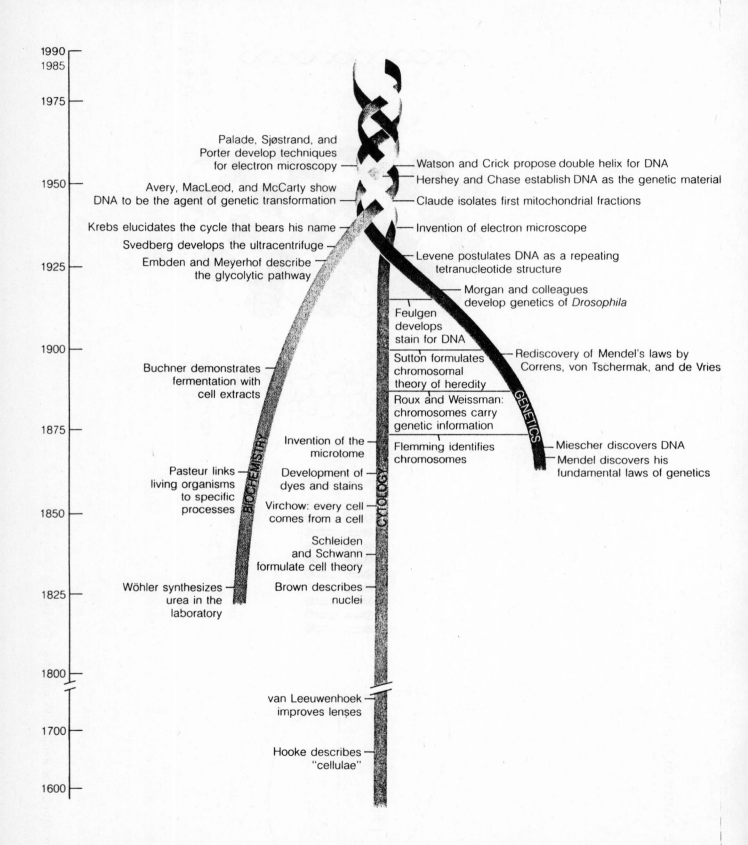

1990
1985
1975
1950
1925
1900
1875
1850
1825
1800
1700
1600

Palade, Sjøstrand, and Porter develop techniques for electron microscopy

Watson and Crick propose double helix for DNA

Avery, MacLeod, and McCarty show DNA to be the agent of genetic transformation

Hershey and Chase establish DNA as the genetic material

Claude isolates first mitochondrial fractions

Krebs elucidates the cycle that bears his name

Invention of electron microscope

Svedberg develops the ultracentrifuge

Embden and Meyerhof describe the glycolytic pathway

Levene postulates DNA as a repeating tetranucleotide structure

Morgan and colleagues develop genetics of *Drosophila*

Feulgen develops stain for DNA

Buchner demonstrates fermentation with cell extracts

Rediscovery of Mendel's laws by Correns, von Tschermak, and de Vries

Sutton formulates chromosomal theory of heredity

Roux and Weissman: chromosomes carry genetic information

GENETICS

Invention of the microtome

Flemming identifies chromosomes

Miescher discovers DNA

Mendel discovers his fundamental laws of genetics

Pasteur links living organisms to specific processes

Development of dyes and stains

BIOCHEMISTRY

Virchow: every cell comes from a cell

Schleiden and Schwann formulate cell theory

CYTOLOGY

Wöhler synthesizes urea in the laboratory

Brown describes nuclei

van Leeuwenhoek improves lenses

Hooke describes "cellulae"

Copyright © The Benjamin/Cummings Publishing Company, Inc.

Energies of Biologically Important Transitions, Bonds, and Wavelengths of Electromagnetic Radiation

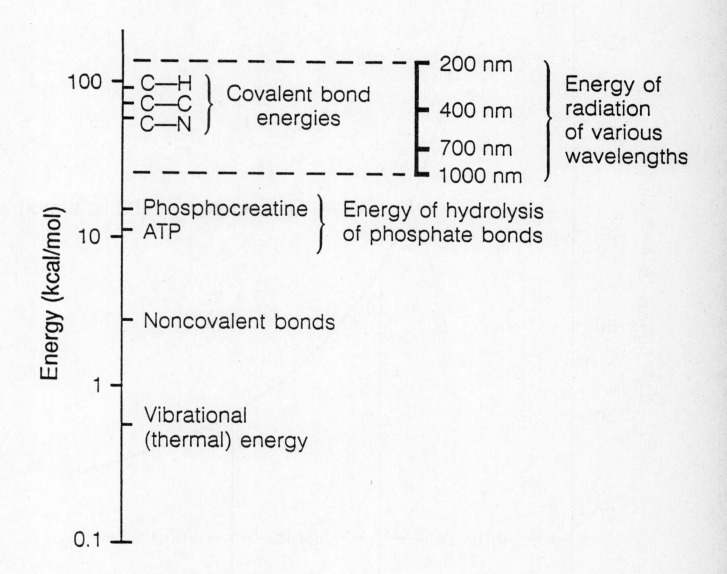

Copyright © The Benjamin/Cummings Publishing Company, Inc.

The Relationship Between Energy and Wavelength for Electromagnetic Radiation

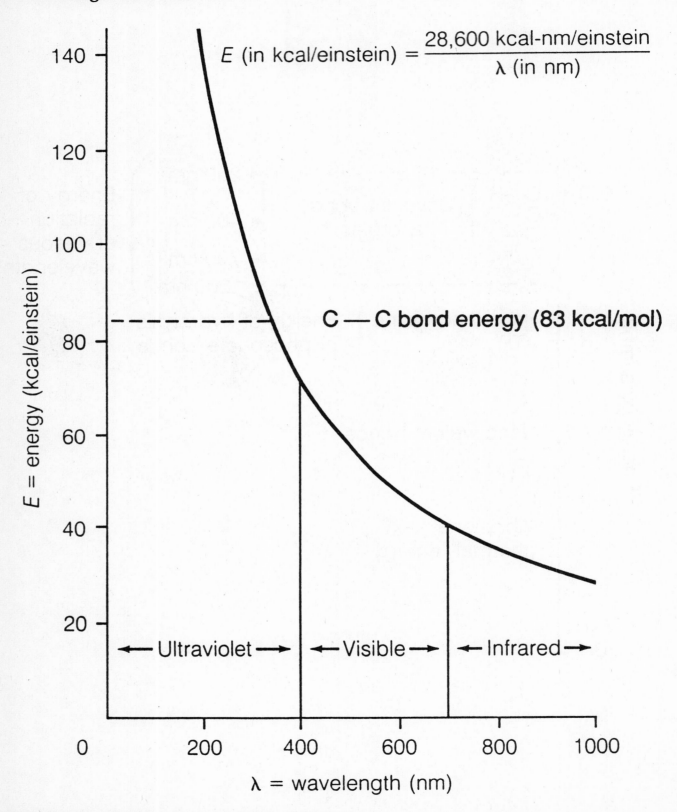

$$E \text{ (in kcal/einstein)} = \frac{28{,}600 \text{ kcal-nm/einstein}}{\lambda \text{ (in nm)}}$$

C — C bond energy (83 kcal/mol)

E = energy (kcal/einstein)

← Ultraviolet → ← Visible → ← Infrared →

λ = wavelength (nm)

Copyright © The Benjamin/Cummings Publishing Company, Inc.

Stereoisomers

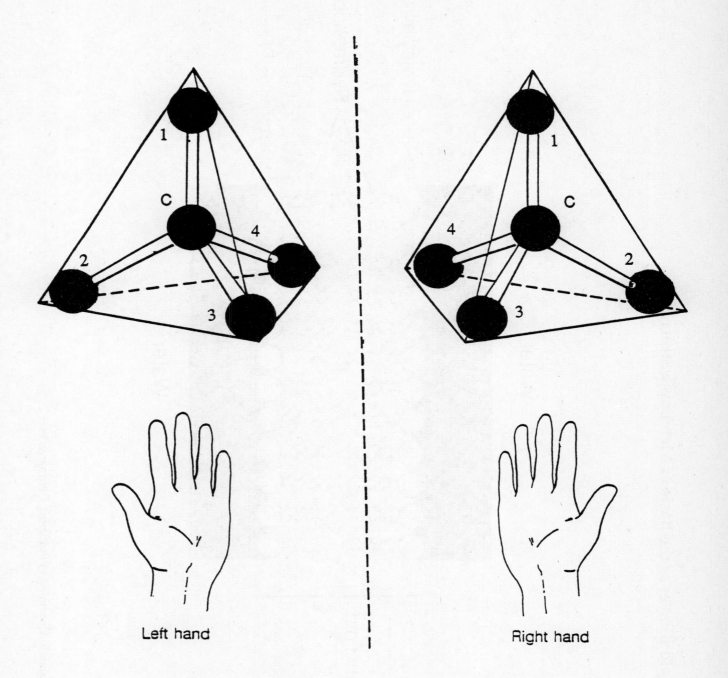

Left hand

Right hand

Copyright © The Benjamin/Cummings Publishing Company, Inc.

The Phospholipid Bilayer as the Basis of Membrane Structure

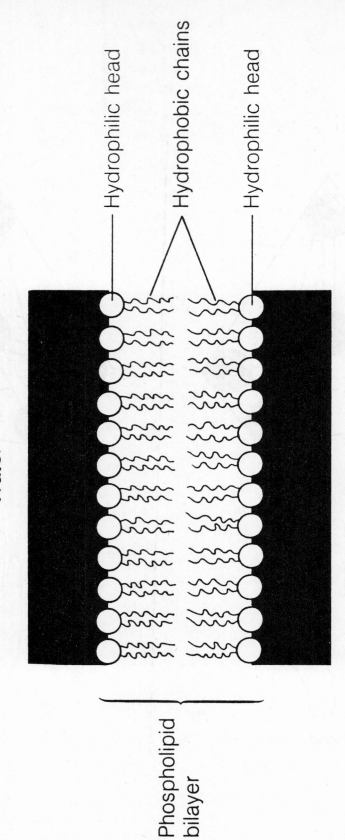

Hydrophilic head

Hydrophobic chains

Hydrophilic head

Water

Water

Phospholipid bilayer

Copyright © The Benjamin/Cummings Publishing Company, Inc.

The Spontaneity of Polypeptide Folding

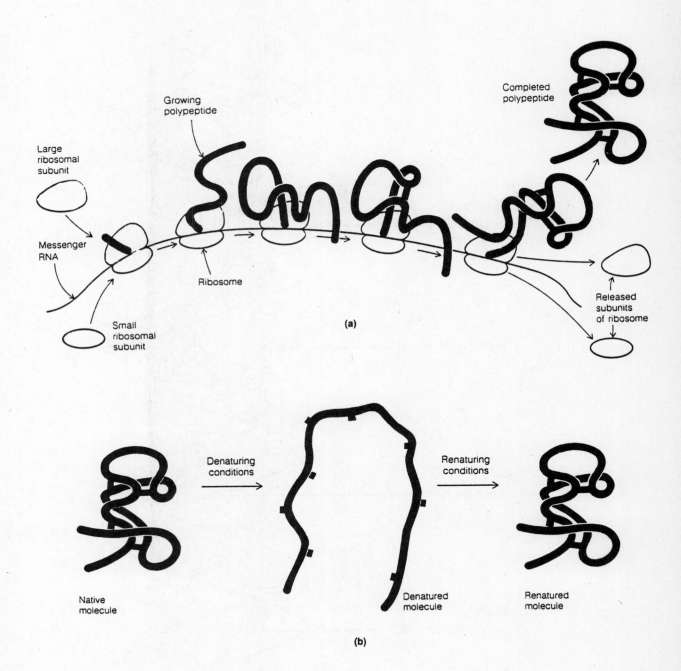

Large
ribosomal
subunit

Growing
polypeptide

Completed
polypeptide

Messenger
RNA

Ribosome

Small
ribosomal
subunit

Released
subunits
of ribosome

(a)

Native
molecule

Denaturing
conditions

Renaturing
conditions

Denatured
molecule

Renatured
molecule

(b)

Copyright © The Benjamin/Cummings Publishing Company, Inc.

The α Helix

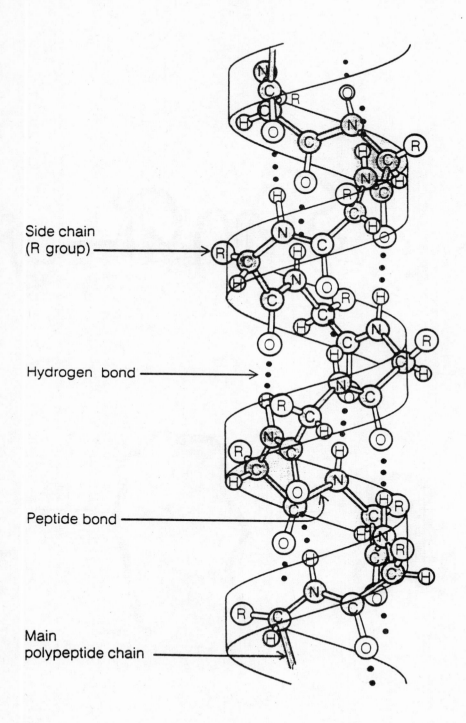

Side chain
(R group)

Hydrogen bond

Peptide bond

Main
polypeptide chain

Copyright © The Benjamin/Cummings Publishing Company, Inc.

The β Pleated Sheet

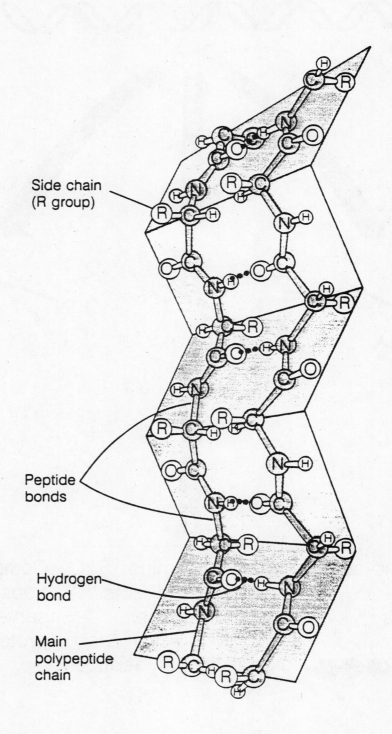

Side chain
(R group)

Peptide
bonds

Hydrogen
bond

Main
polypeptide
chain

Copyright © The Benjamin/Cummings Publishing Company, Inc.

The Roles of DNA and RNA in Protein Synthesis

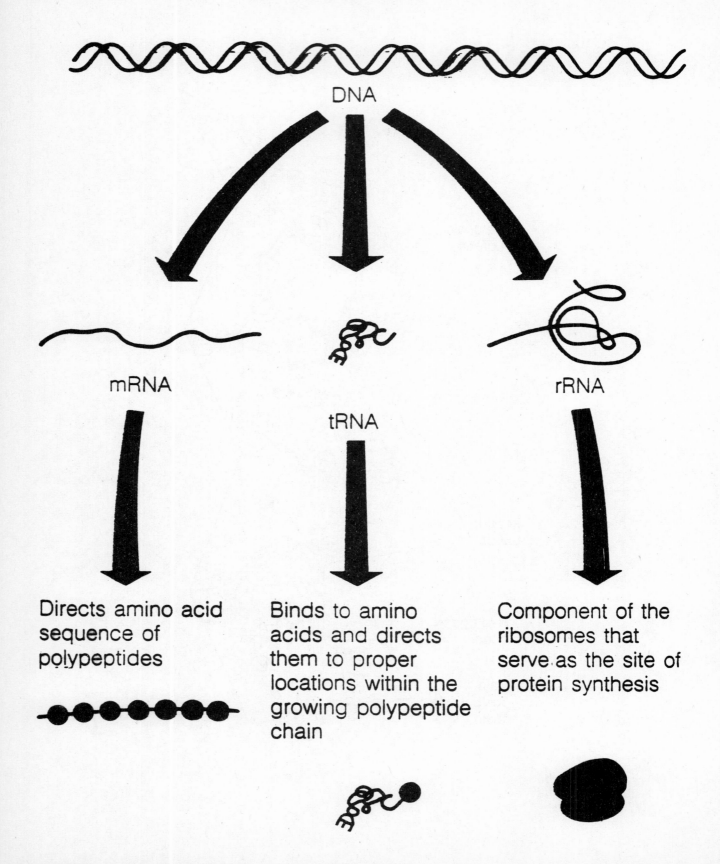

DNA

mRNA

tRNA

rRNA

Directs amino acid sequence of polypeptides

Binds to amino acids and directs them to proper locations within the growing polypeptide chain

Component of the ribosomes that serve as the site of protein synthesis

Copyright © The Benjamin/Cummings Publishing Company, Inc.

Storage of Genetic Information by Prokaryotic and Eukaryotic Cells

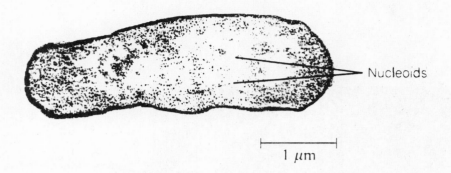

Nucleoids

1 μm

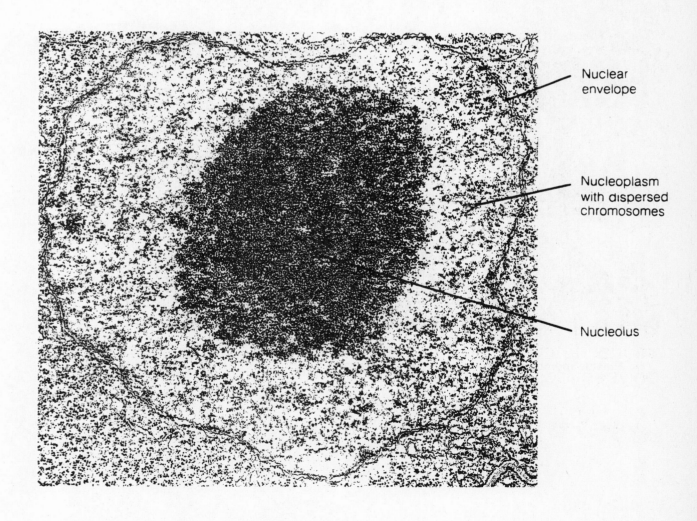

Nuclear envelope

Nucleoplasm with dispersed chromosomes

Nucleolus

Copyright © The Benjamin/Cummings Publishing Company, Inc.

Structural Features of a Eukaryotic Cell

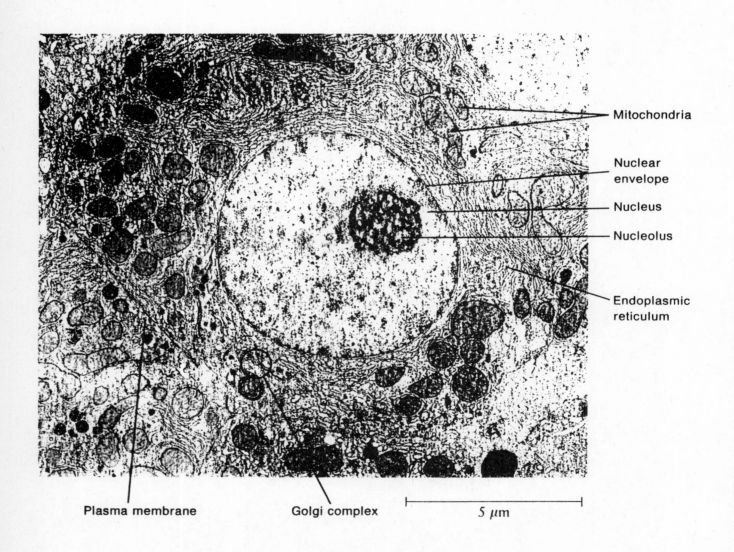

Mitochondria

Nuclear envelope

Nucleus

Nucleolus

Endoplasmic reticulum

Plasma membrane

Golgi complex

5 μm

Copyright © The Benjamin/Cummings Publishing Company, Inc.

A Typical Animal Cell

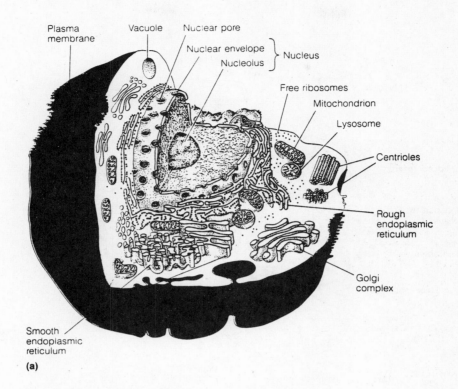

Plasma membrane · Vacuole · Nuclear pore · Nuclear envelope · Nucleolus · Nucleus · Free ribosomes · Mitochondrion · Lysosome · Centrioles · Rough endoplasmic reticulum · Golgi complex · Smooth endoplasmic reticulum

(a)

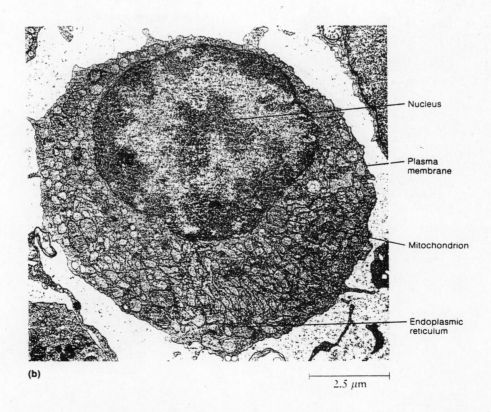

Nucleus · Plasma membrane · Mitochondrion · Endoplasmic reticulum

(b)

2.5 μm

Copyright © The Benjamin/Cummings Publishing Company, Inc.

A Typical Plant Cell

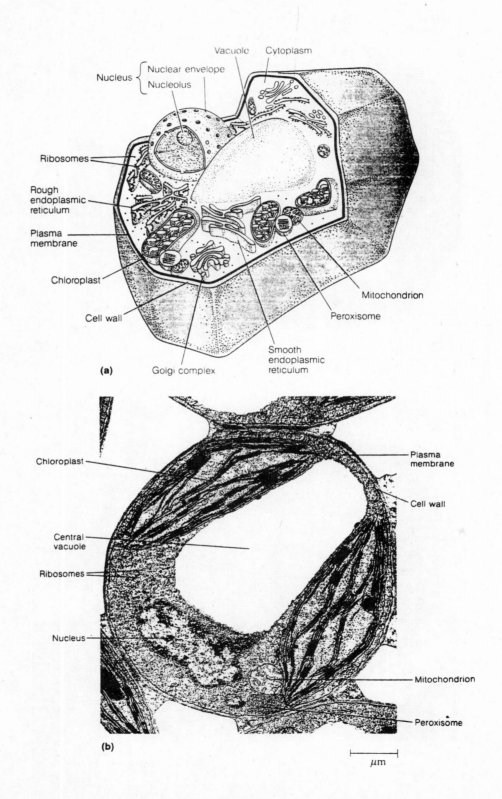

(a)

Nucleus { Nuclear envelope / Nucleolus

Vacuole Cytoplasm

Ribosomes

Rough endoplasmic reticulum

Plasma membrane

Chloroplast

Cell wall

Golgi complex

Smooth endoplasmic reticulum

Peroxisome

Mitochondrion

(b)

Chloroplast

Central vacuole

Ribosomes

Nucleus

Plasma membrane

Cell wall

Mitochondrion

Peroxisome

μm

Copyright © The Benjamin/Cummings Publishing Company, Inc.

Organization of the Plasma Membrane

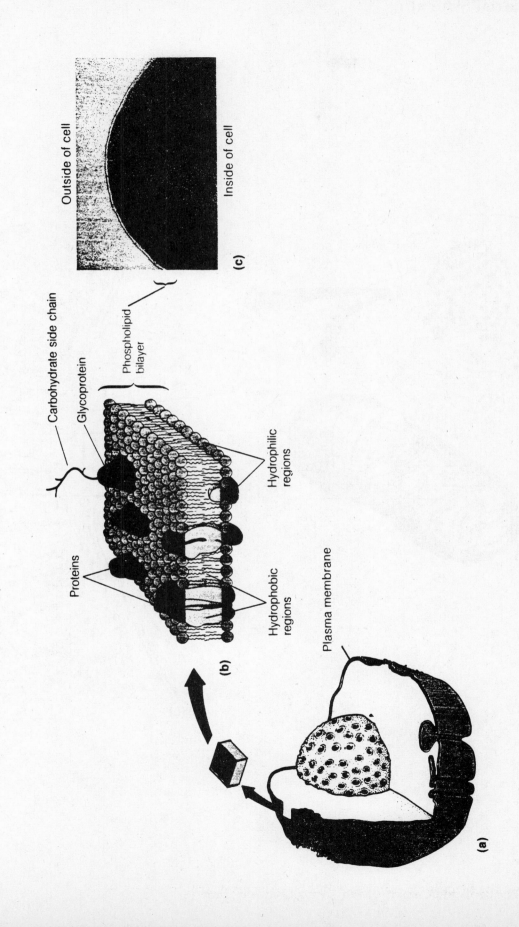

Outside of cell

Inside of cell

(c)

Carbohydrate side chain

Glycoprotein

Phospholipid bilayer

Proteins

Hydrophilic regions

Hydrophobic regions

(b)

Plasma membrane

(a)

Copyright © The Benjamin/Cummings Publishing Company, Inc.

Mitochondrial Structure

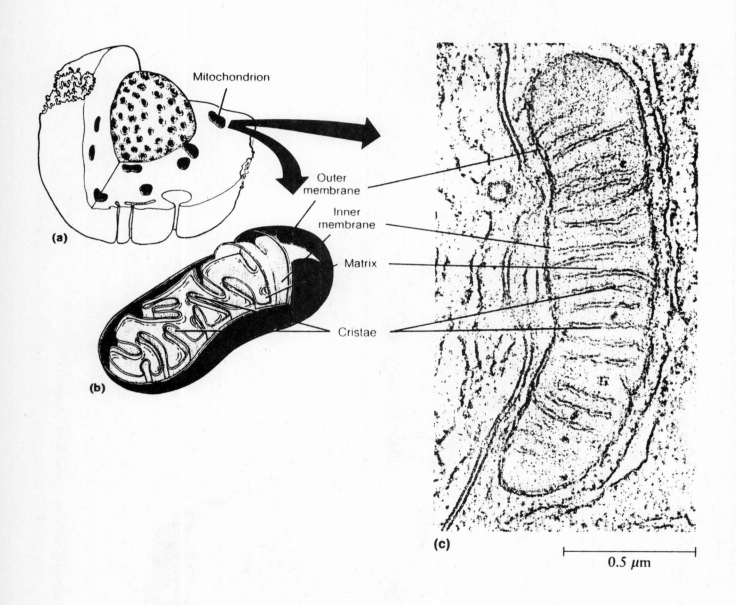

Mitochondrion

Outer
membrane

Inner
membrane

Matrix

Cristae

(a)

(b)

(c)

0.5 μm

Copyright © The Benjamin/Cummings Publishing Company, Inc.

Chloroplast Structure

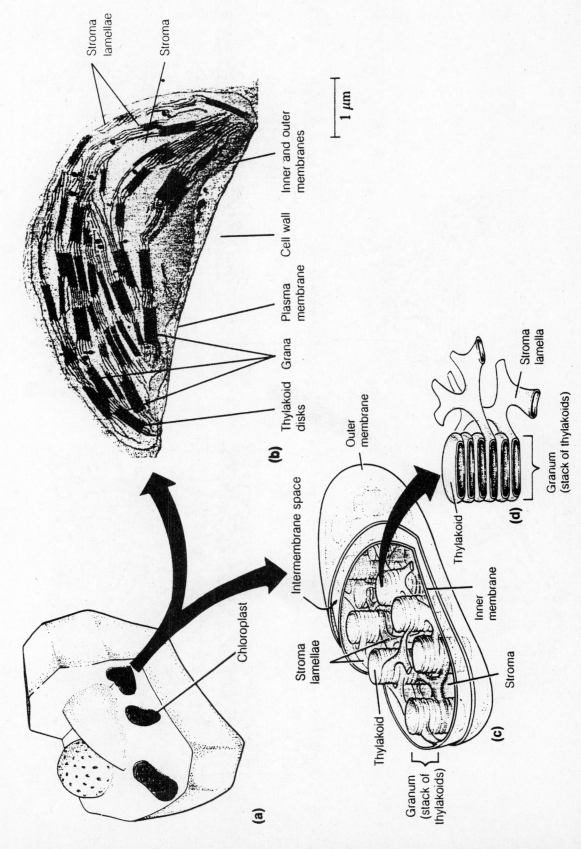

(a)

Chloroplast

(b)

Stroma lamellae

Stroma

Inner and outer membranes

Cell wall

Plasma membrane

Grana

Thylakoid disks

1 μm

(c)

Intermembrane space

Outer membrane

Stroma lamellae

Thylakoid

Granum (stack of thylakoids)

Inner membrane

Stroma

(d)

Stroma lamella

Thylakoid

Granum (stack of thylakoids)

Copyright © The Benjamin/Cummings Publishing Company, Inc.

The Endoplasmic Reticulum

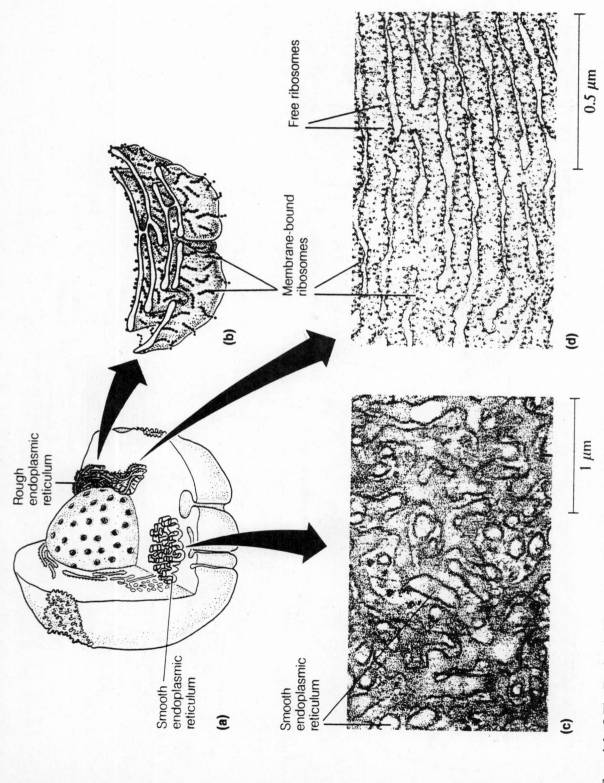

Rough endoplasmic reticulum

Smooth endoplasmic reticulum

(a)

(b)

Membrane-bound ribosomes

Free ribosomes

0.5 μm

(d)

Smooth endoplasmic reticulum

1 μm

(c)

Copyright © The Benjamin/Cummings Publishing Company, Inc.

The Golgi Complex

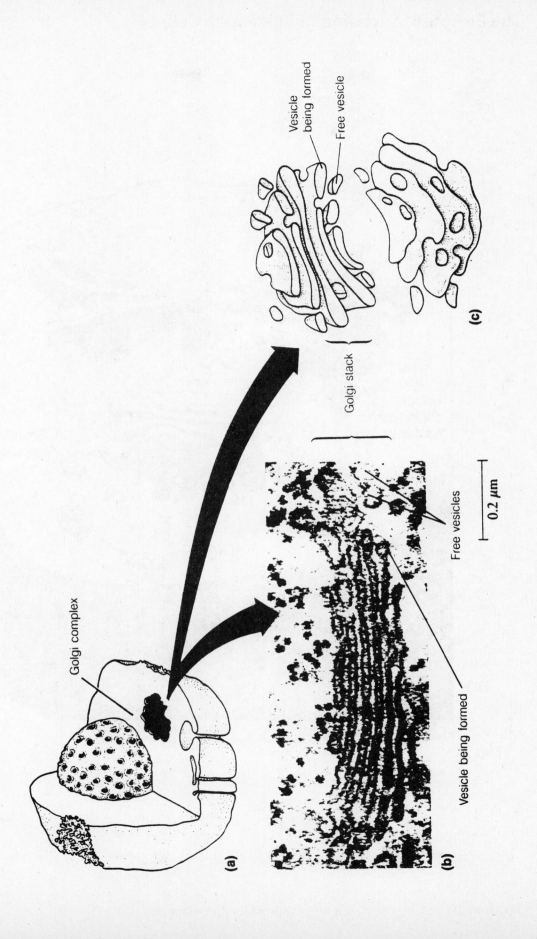

Golgi complex

(a)

Vesicle being formed

Free vesicles

0.2 μm

(b)

Vesicle being formed

Free vesicle

Golgi stack

(c)

Copyright © The Benjamin/Cummings Publishing Company, Inc.

The Process of Secretion in Eukaryotic Cells

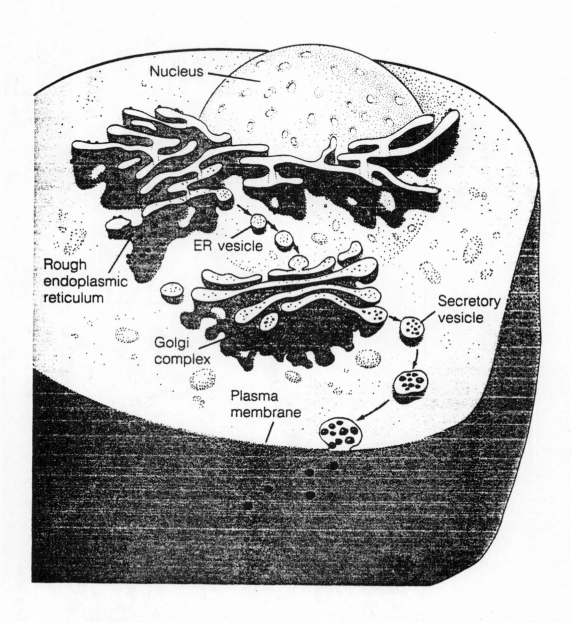

Nucleus

ER vesicle

Rough
endoplasmic
reticulum

Secretory
vesicle

Golgi
complex

Plasma
membrane

Copyright © The Benjamin/Cummings Publishing Company, Inc.

Structures of Microtubules, Microfilaments, and Intermediate Filaments

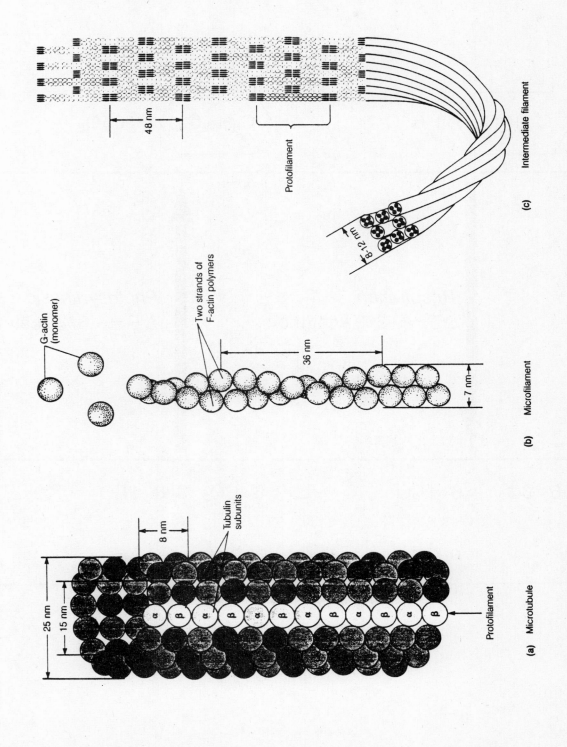

(a) Microtubule

8 nm

25 nm

15 nm

Tubulin subunits

Protofilament

α β α β α β α β α β α β

(b) Microfilament

G-actin (monomer)

Two strands of F-actin polymers

36 nm

7 nm

(c) Intermediate filament

48 nm

Protofilament

8–12 nm

Copyright © The Benjamin/Cummings Publishing Company, Inc.

Internal Energy Changes

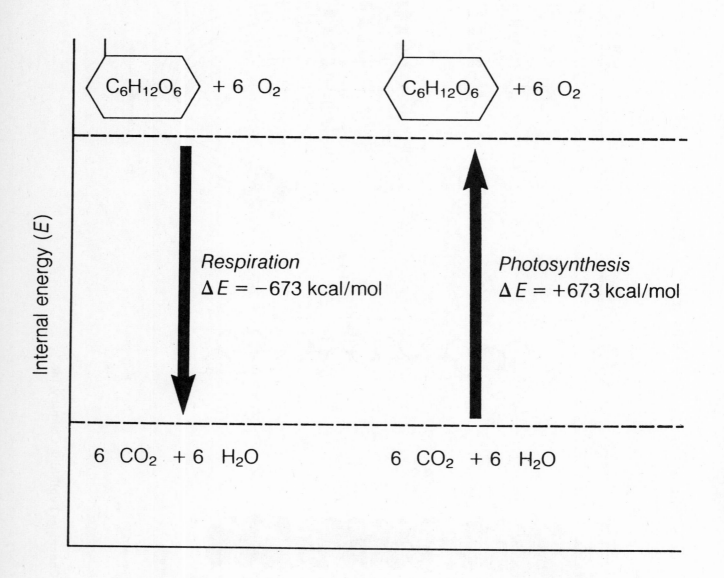

Copyright © The Benjamin/Cummings Publishing Company, Inc.

Free Energy and Chemical Equilibrium

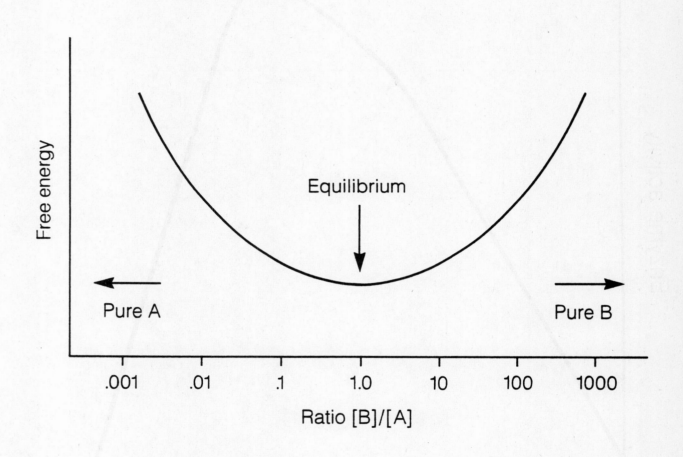

Copyright © The Benjamin/Cummings Publishing Company, Inc.

Effect of Temperature on the Activity of an Enzyme

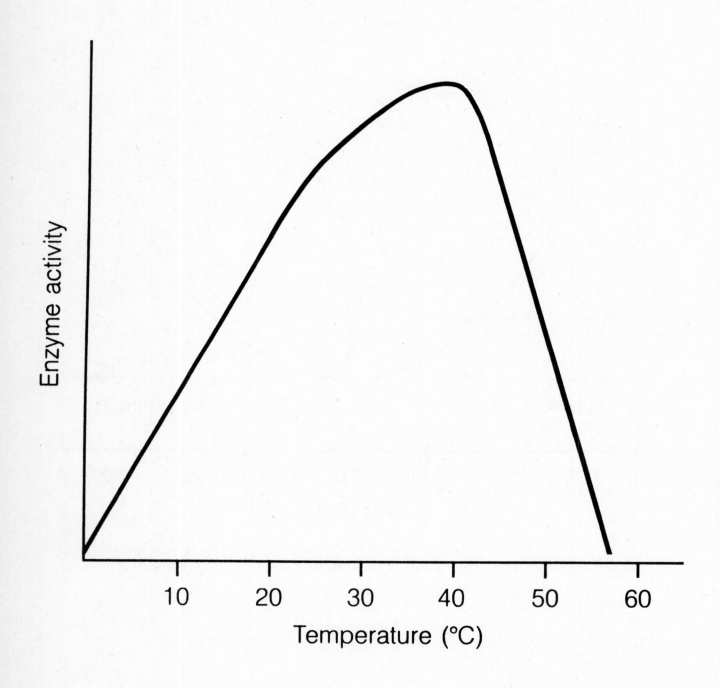

Copyright © The Benjamin/Cummings Publishing Company, Inc.

Effect of pH on the Activity of an Enzyme

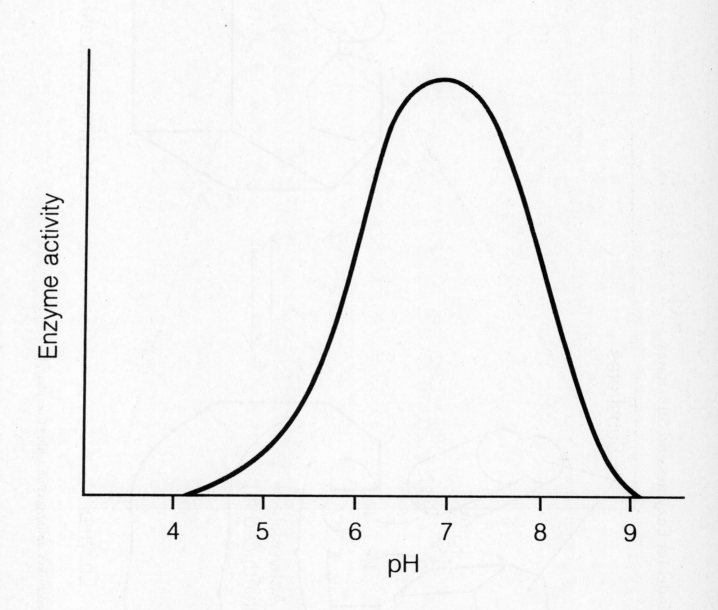

Copyright © The Benjamin/Cummings Publishing Company, Inc.

Induced-Fit Model for Enzymes and Substrates

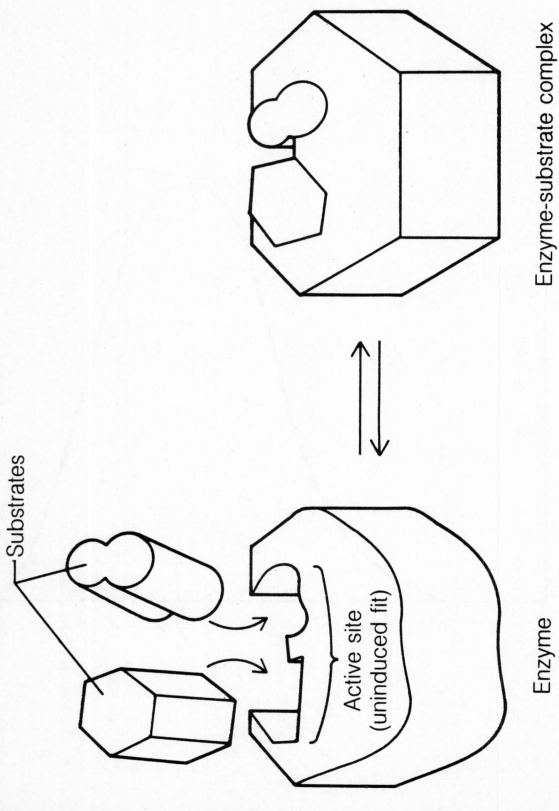

Substrates

Active site
(uninduced fit)

Enzyme

Enzyme-substrate complex

Copyright © The Benjamin/Cummings Publishing Company, Inc.

Relationship Between Reaction Velocity and Substrate Concentration

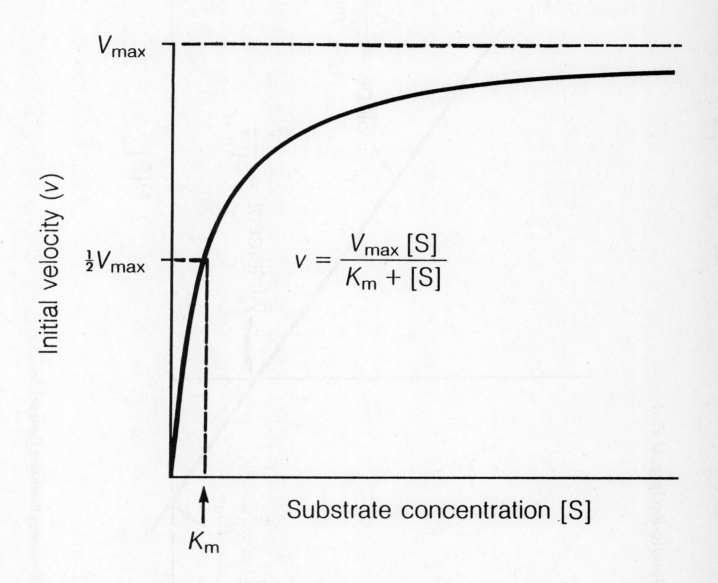

$$v = \frac{V_{max}\,[S]}{K_m + [S]}$$

Copyright © The Benjamin/Cummings Publishing Company, Inc.

Lineweaver-Burk Double-Reciprocal Plot

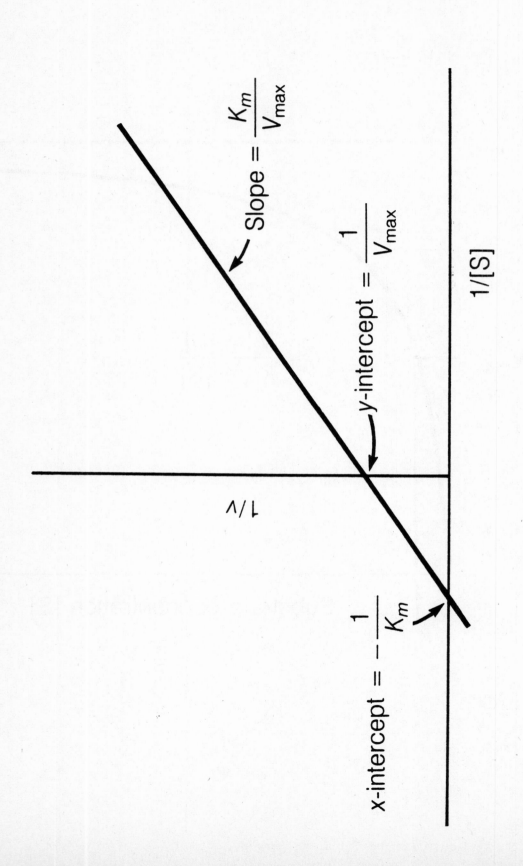

Copyright © The Benjamin/Cummings Publishing Company, Inc.

Effect of Inhibition on the Double-Reciprocal Plot

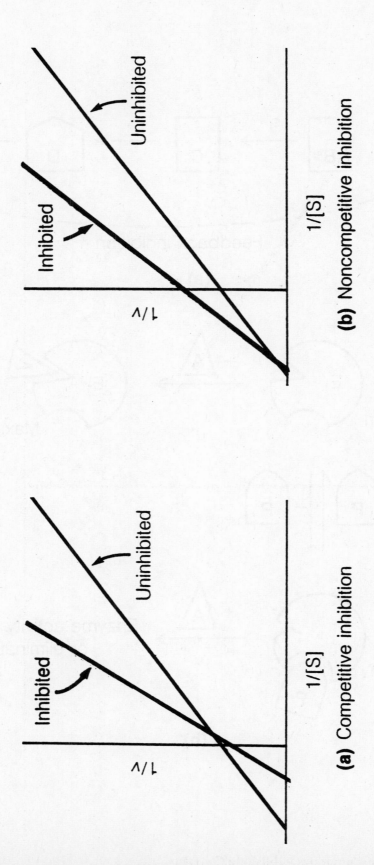

Copyright © The Benjamin/Cummings Publishing Company, Inc.

Allosteric Regulation of Enzyme Activity

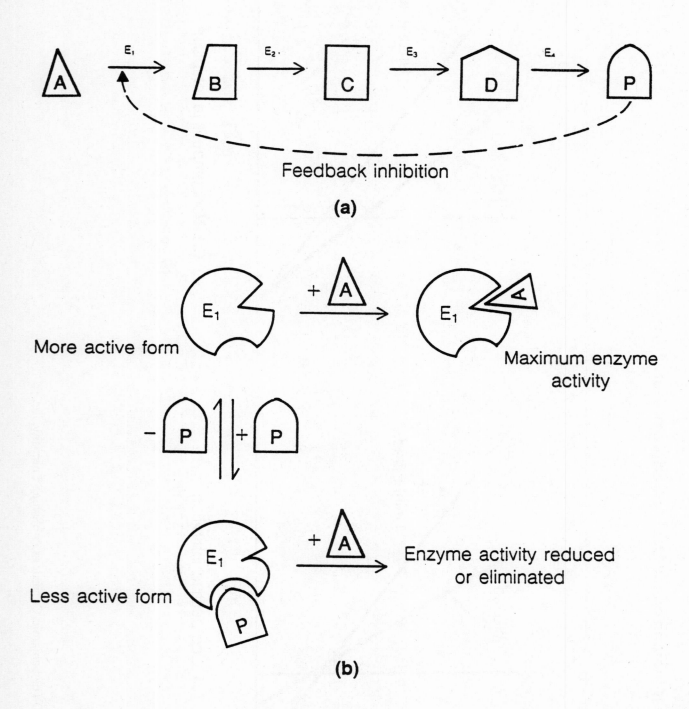

Feedback inhibition

(a)

More active form + A → Maximum enzyme activity

− P ⇌ + P

Less active form + A → Enzyme activity reduced or eliminated

(b)

Copyright © The Benjamin/Cummings Publishing Company, Inc.

Time Line for the Development of Our Understanding of Membrane Chemistry and Structure

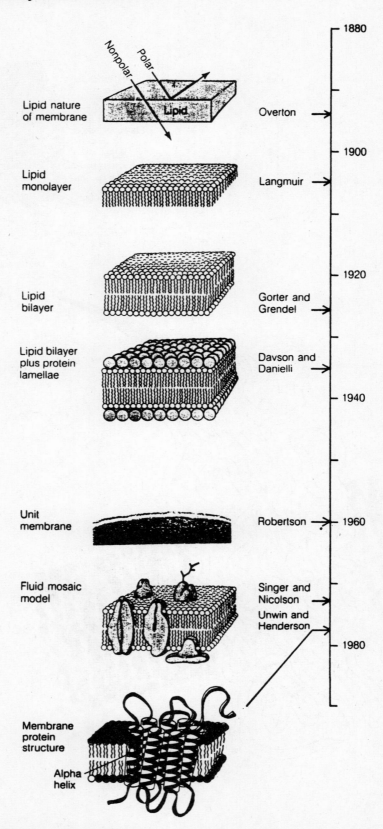

Lipid nature of membrane — Overton — 1880

Lipid monolayer — Langmuir — 1900

Lipid bilayer — Gorter and Grendel — 1920

Lipid bilayer plus protein lamellae — Davson and Danielli — 1940

Unit membrane — Robertson — 1960

Fluid mosaic model — Singer and Nicolson

Unwin and Henderson — 1980

Membrane protein structure

Alpha helix

Nonpolar   Polar

Lipid

Copyright © The Benjamin/Cummings Publishing Company, Inc.

Fluid Mosaic Model of Membrane Structure

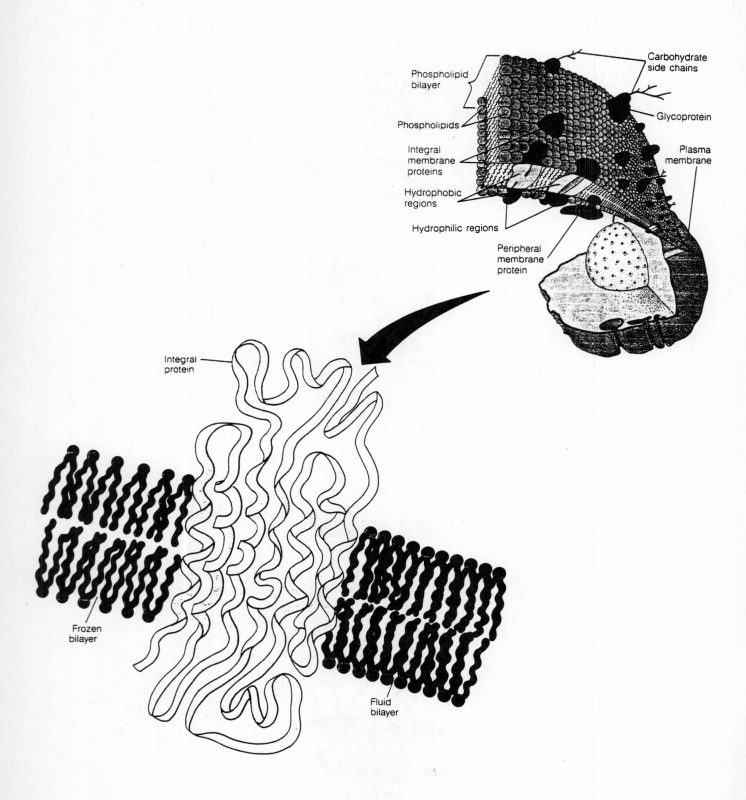

Copyright © The Benjamin/Cummings Publishing Company, Inc.

Structure of an Integral Protein

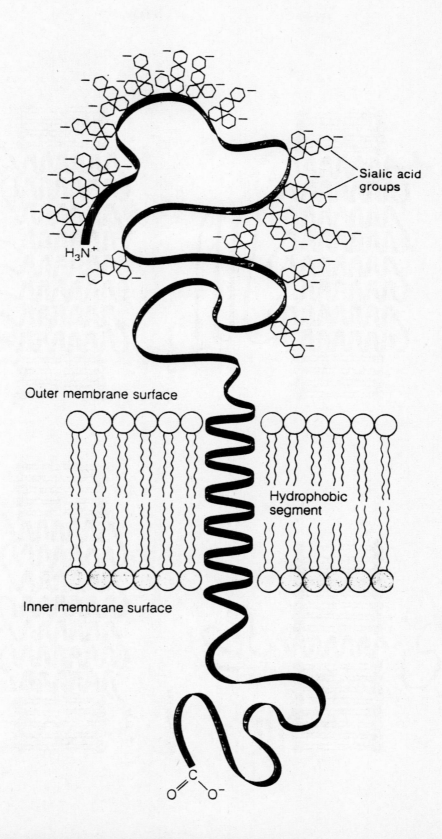

Sialic acid groups

H₃N⁺

Outer membrane surface

Hydrophobic segment

Inner membrane surface

Copyright © The Benjamin/Cummings Publishing Company, Inc.

Multiple Transmembrane Sequences in Integral Proteins

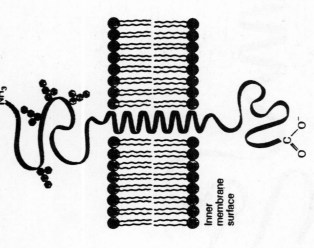

Inner membrane surface

(a) Glycophorin

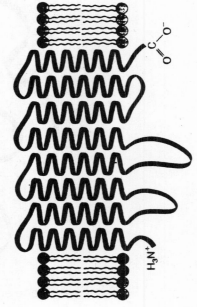

H_3N^+

(b) Anion channel

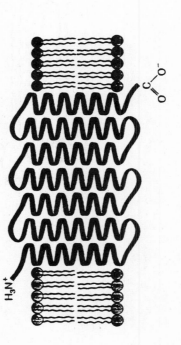

H_3N^+

(c) Bacteriorhodopsin

H_3N^+

(d) Calcium transport ATPase

Copyright © The Benjamin/Cummings Publishing Company, Inc.

Structure of the Red Cell Membrane

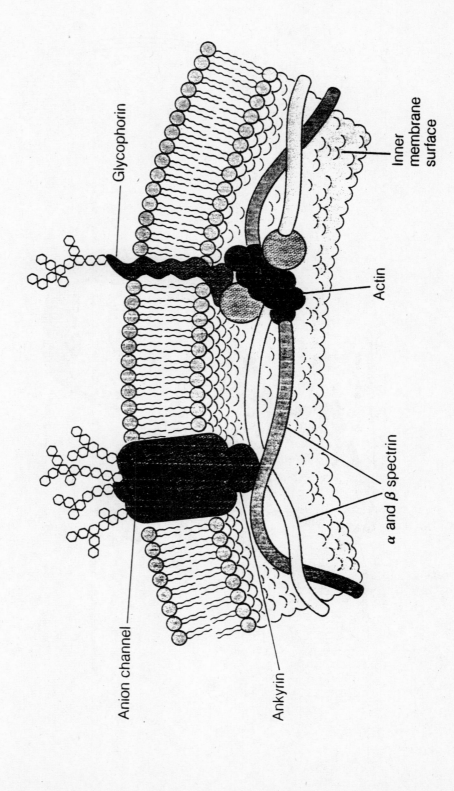

Glycophorin

Anion channel

Ankyrin

Actin

α and β spectrin

Inner membrane surface

Copyright © The Benjamin/Cummings Publishing Company, Inc.

Freeze-Fracturing a Membrane

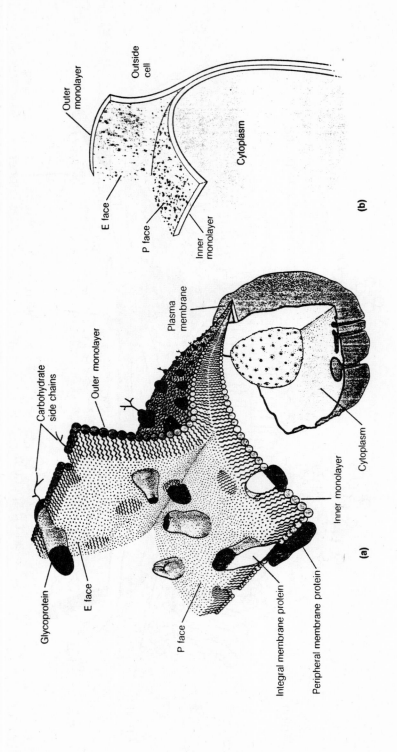

Carbohydrate side chains

Outer monolayer

Plasma membrane

Glycoprotein

E face

P face

Integral membrane protein

Peripheral membrane protein

Inner monolayer

Cytoplasm

(a)

Outer monolayer

Outside cell

E face

P face

Inner monolayer

Cytoplasm

(b)

Copyright © The Benjamin/Cummings Publishing Company, Inc.

Patching and Capping as a Demonstration of the Mobility of Membrane Proteins

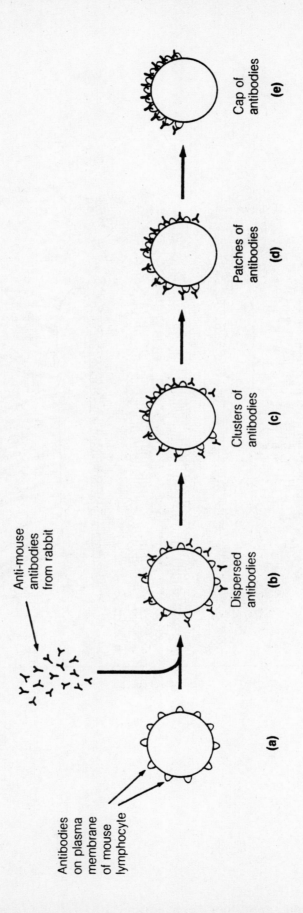

Antibodies on plasma membrane of mouse lymphocyte

Anti-mouse antibodies from rabbit

(a)

(b) Dispersed antibodies

(c) Clusters of antibodies

(d) Patches of antibodies

(e) Cap of antibodies

Copyright © The Benjamin/Cummings Publishing Company, Inc.

The Desmosome

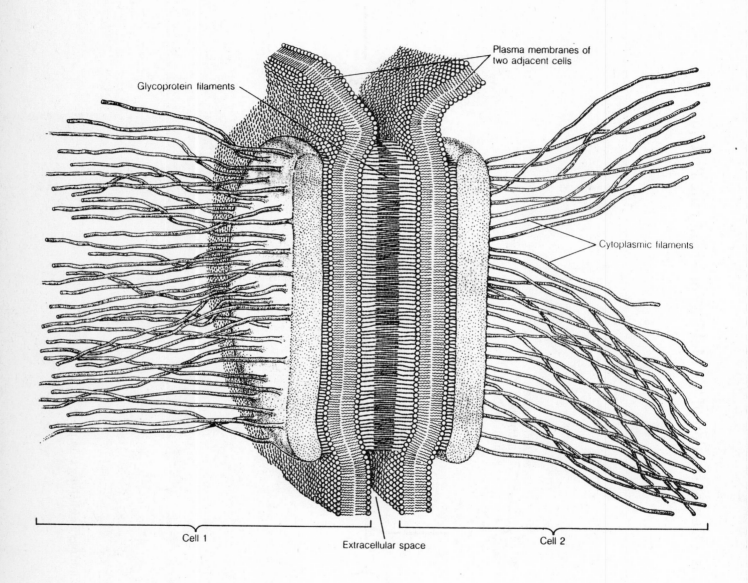

Glycoprotein filaments

Plasma membranes of two adjacent cells

Cytoplasmic filaments

Cell 1

Extracellular space

Cell 2

Copyright © The Benjamin/Cummings Publishing Company, Inc.

The Tight Junction

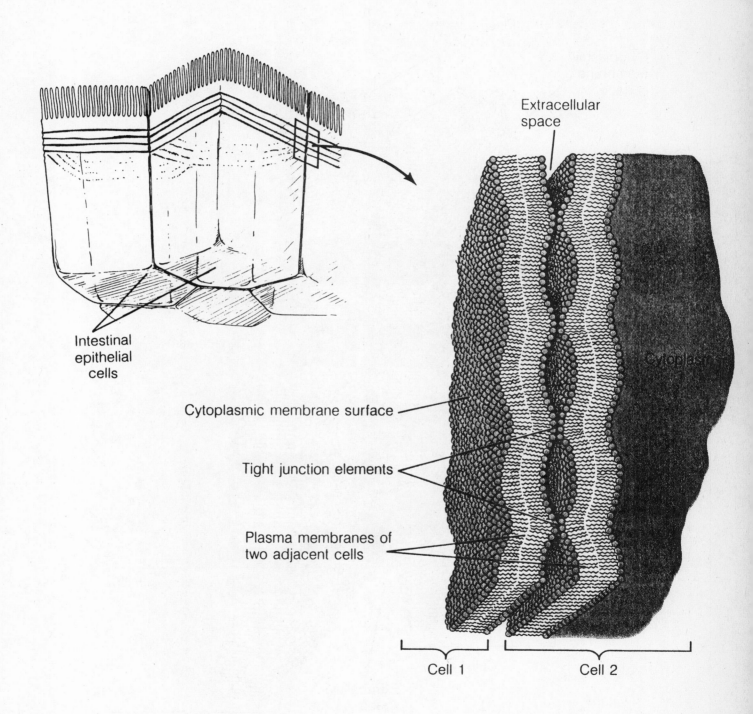

Intestinal epithelial cells

Extracellular space

Cytoplasmic membrane surface

Tight junction elements

Plasma membranes of two adjacent cells

Cytoplasm

Cell 1

Cell 2

Copyright © The Benjamin/Cummings Publishing Company, Inc.

The Gap Junction

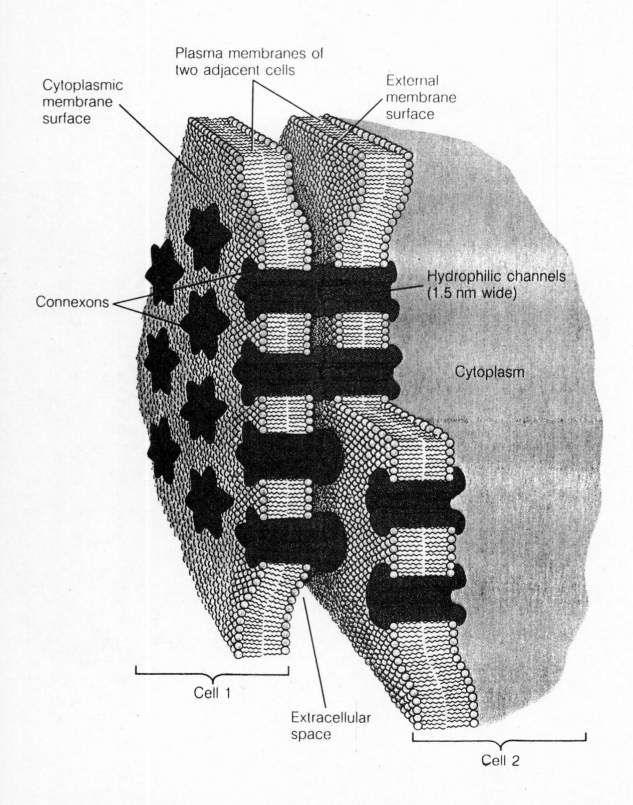

Cytoplasmic membrane surface

Plasma membranes of two adjacent cells

External membrane surface

Connexons

Hydrophilic channels (1.5 nm wide)

Cytoplasm

Cell 1

Extracellular space

Cell 2

Copyright © The Benjamin/Cummings Publishing Company, Inc.

Transport Processes Operative Within a Composite Eukaryotic Cell

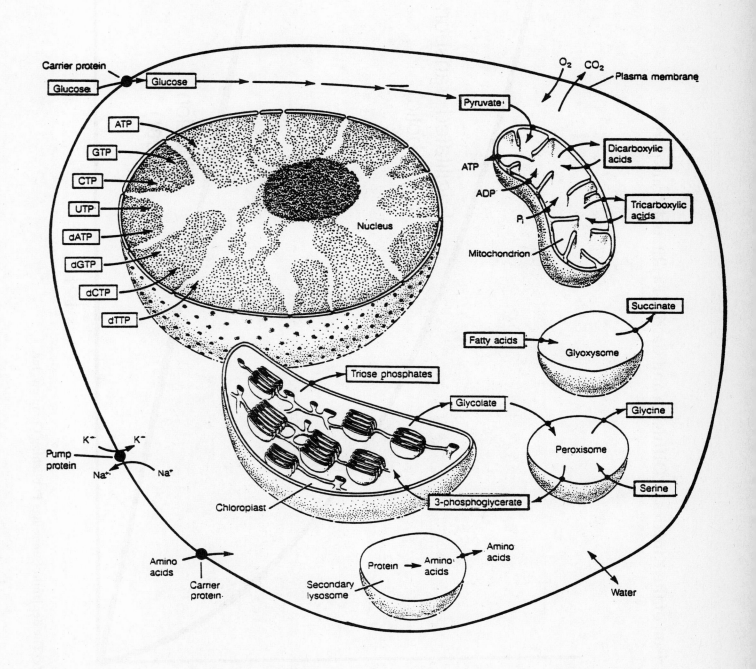

Copyright © The Benjamin/Cummings Publishing Company, Inc.

Comparison of the Kinetics of Diffusion and Carrier-Facilitated Transport

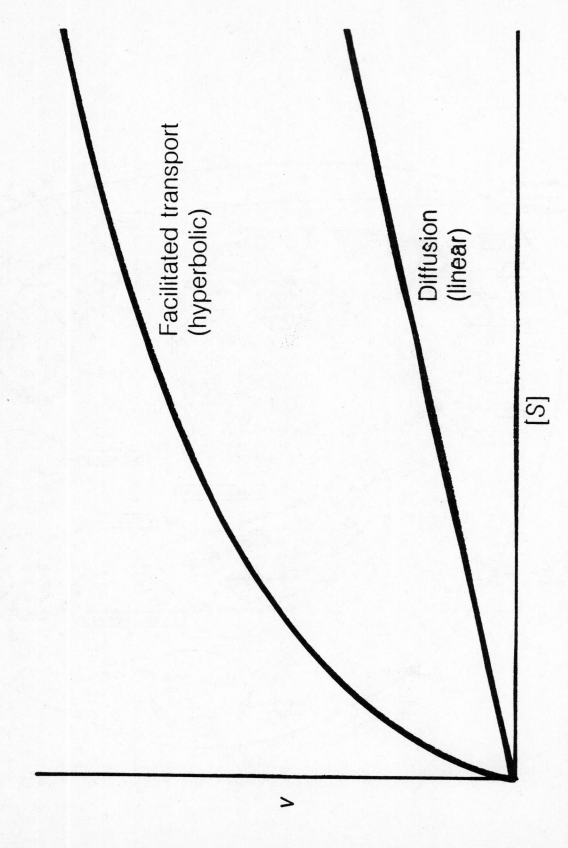

Copyright © The Benjamin/Cummings Publishing Company, Inc.

Possible Mechanism for Facilitated Transport

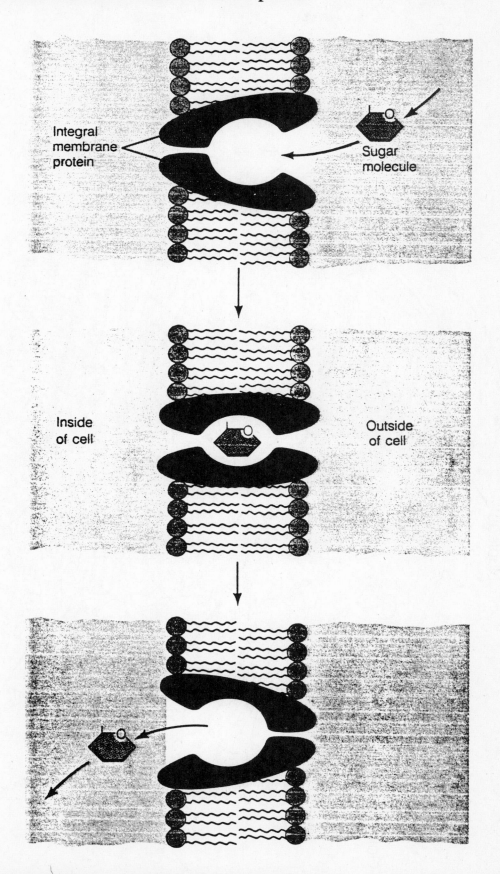

Integral membrane protein

Sugar molecule

Inside of cell

Outside of cell

Copyright © The Benjamin/Cummings Publishing Company, Inc.

Smooth and Rough ER and Their Relation to the Nuclear Envelope and the Golgi Complex

Rought ER

Nucleus

Golgi complex

Smooth ER

Copyright © The Benjamin/Cummings Publishing Company, Inc.

Illustration of Golgi Structure

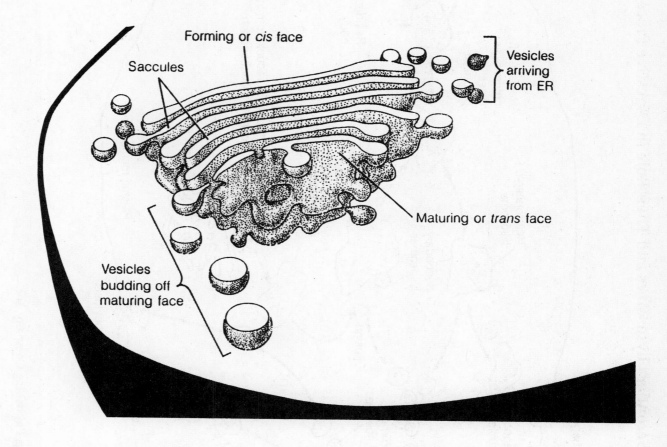

Forming or *cis* face

Saccules

Vesicles arriving from ER

Maturing or *trans* face

Vesicles budding off maturing face

Copyright © The Benjamin/Cummings Publishing Company, Inc.

Formation of Primary and Secondary Lysosomes and Their Role in Digestive Processes

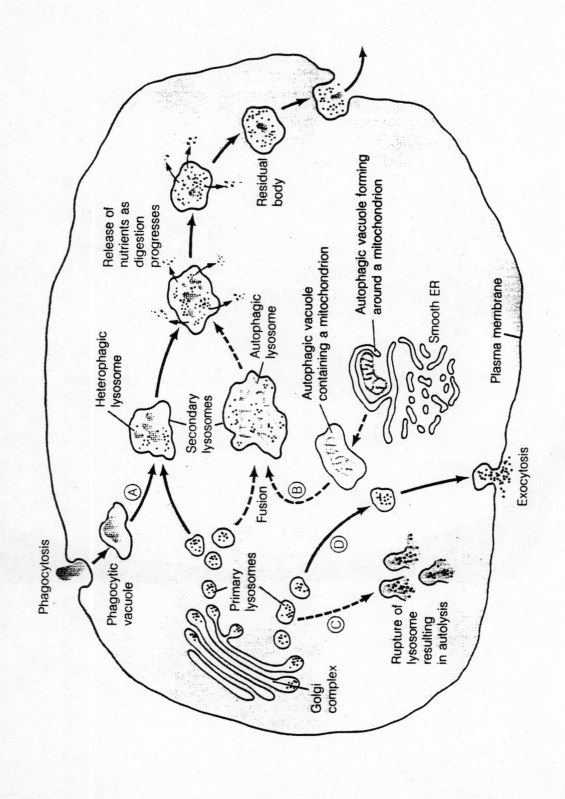

Copyright © The Benjamin/Cummings Publishing Company, Inc.

Standard Free Energies of Hydrolysis for Some Common Phosphorylated Compounds Found in Cells

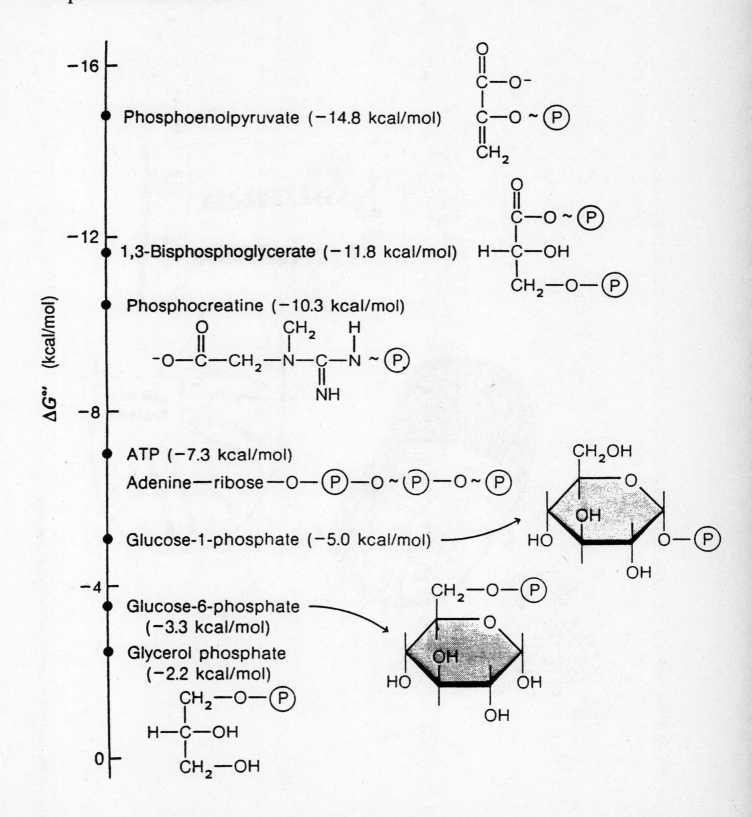

Copyright © The Benjamin/Cummings Publishing Company, Inc.

Mitochondrial Structure

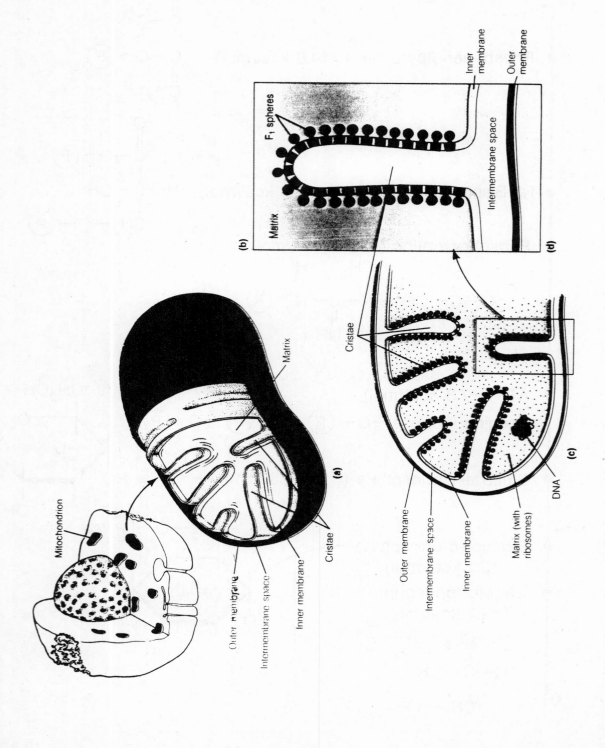

Mitochondrion

Matrix

Outer membrane

Intermembrane space

Inner membrane

Cristae

(a)

F₁ spheres

Matrix

Inner membrane

Outer membrane

Intermembrane space

(b)

(d)

Cristae

Outer membrane

Intermembrane space

Inner membrane

Matrix (with ribosomes)

DNA

(c)

Copyright © The Benjamin/Cummings Publishing Company, Inc.

Energetics of Electron Transport

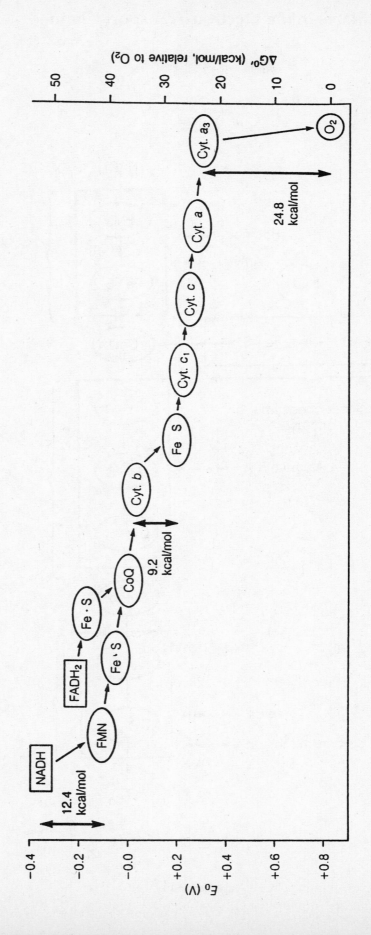

Copyright © The Benjamin/Cummings Publishing Company, Inc.

Order of Intermediates in the Electron Transport Chain

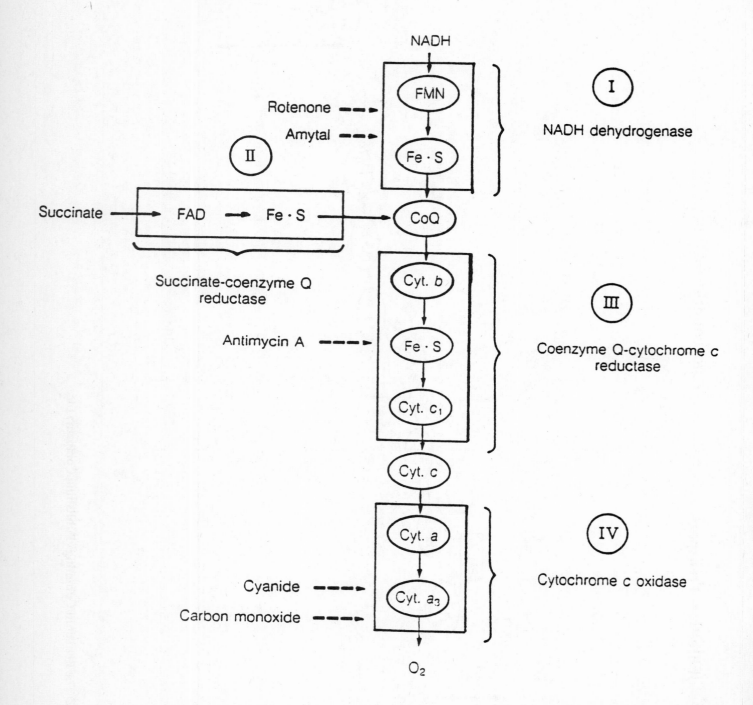

Copyright © The Benjamin/Cummings Publishing Company, Inc.

Vectorial Pumping of Protons Across the Inner Mitochondrial Membrane

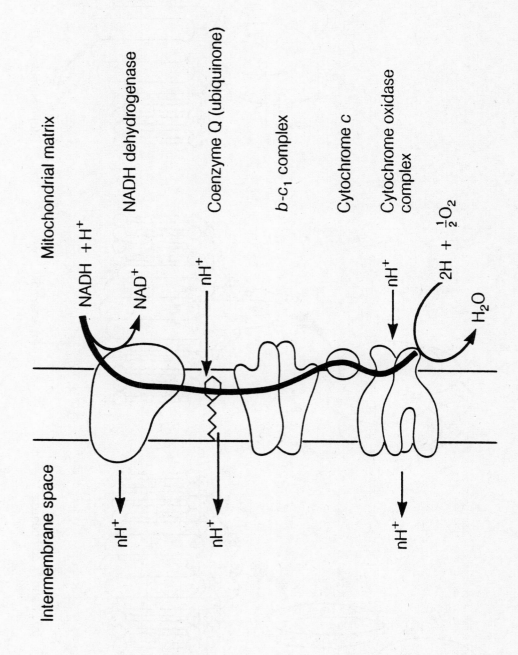

Copyright © The Benjamin/Cummings Publishing Company, Inc.

F_1 and F_0 Components of the Mitochondrial ATP-Synthesizing Enzyme

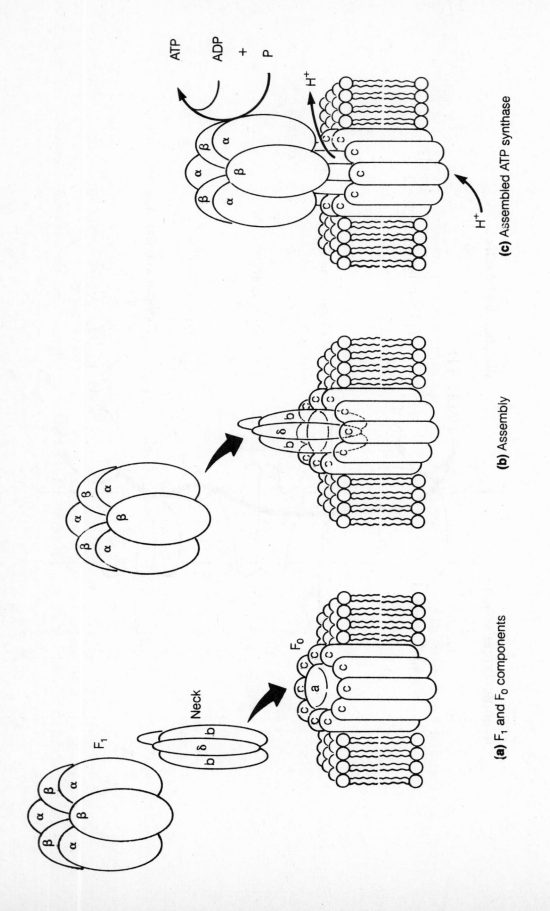

(a) F_1 and F_0 components

(b) Assembly

(c) Assembled ATP synthase

Copyright © The Benjamin/Cummings Publishing Company, Inc.

Dynamics of the Electrochemical Proton Gradient

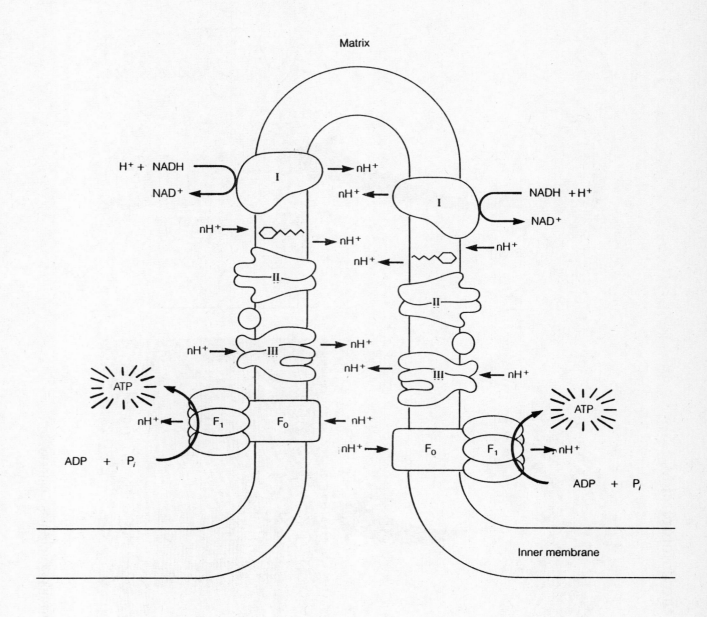

Copyright © The Benjamin/Cummings Publishing Company, Inc.

Structural Features of the Chloroplast

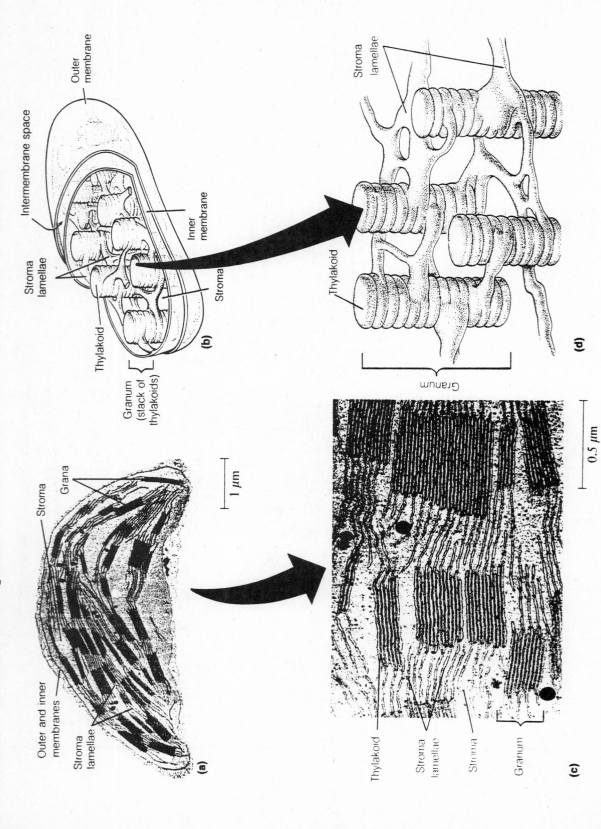

(a)

Grana

Stroma

Outer and inner membranes

Stroma lamellae

1 μm

(b)

Outer membrane

Intermembrane space

Inner membrane

Stroma lamellae

Thylakoid

Stroma

Granum (stack of thylakoids)

(c)

Thylakoid

Stroma lamellae

Stroma

Granum

0.5 μm

(d)

Stroma lamellae

Thylakoid

Granum

Copyright © The Benjamin/Cummings Publishing Company, Inc.

Absorption Spectra for Ether Extracts of Chlorophylls *a* and *b*

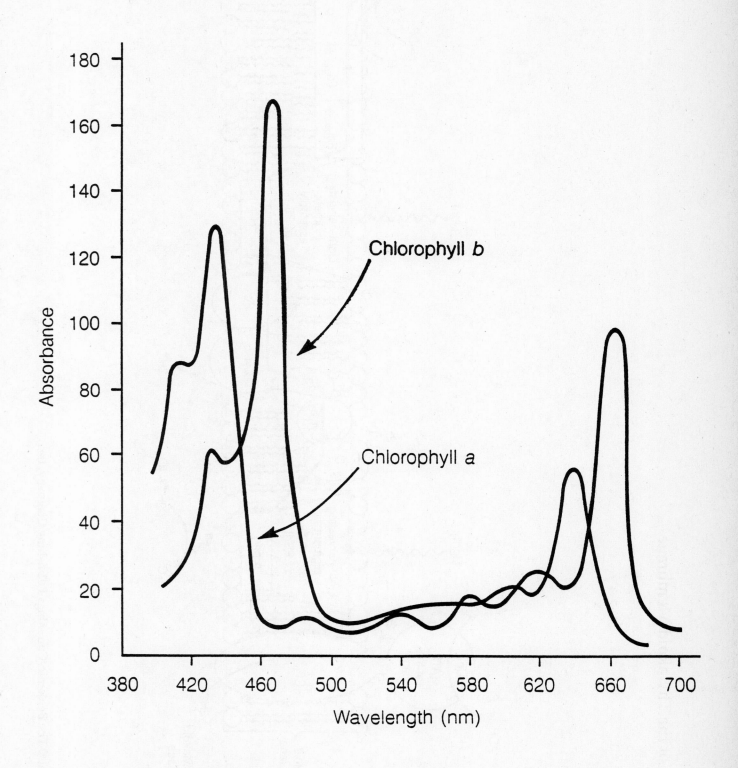

Copyright © The Benjamin/Cummings Publishing Company, Inc.

Model of the Thylakoid Membrane

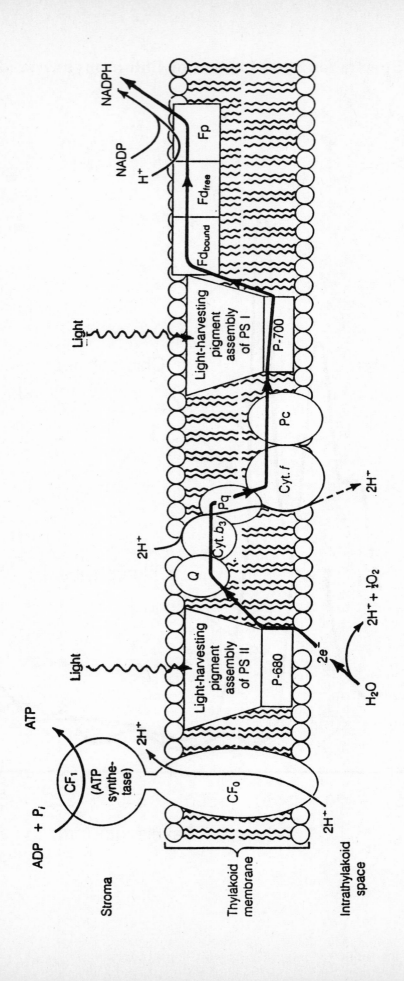

Copyright © The Benjamin/Cummings Publishing Company, Inc.

Localization of the Hatch-Slack and Calvin Cycles Within a C$_4$ Leaf

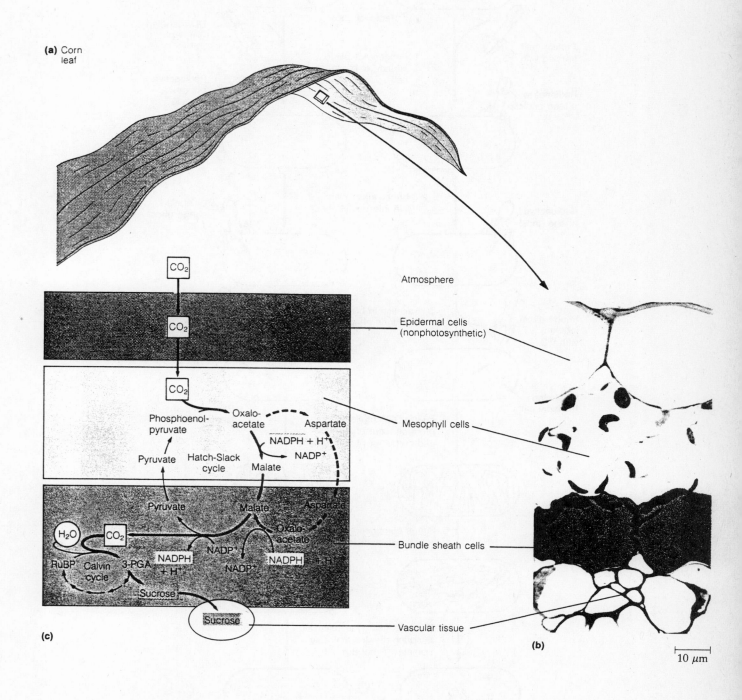

(a) Corn leaf

Atmosphere

Epidermal cells (nonphotosynthetic)

Mesophyll cells

Bundle sheath cells

Vascular tissue

CO$_2$

CO$_2$

CO$_2$

Phosphoenol-pyruvate

Oxalo-acetate

Aspartate

NADPH + H$^+$

NADP$^+$

Pyruvate

Hatch-Slack cycle

Malate

Pyruvate

Malate

Aspartate

H$_2$O

CO$_2$

Oxalo-acetate

RuBP

Calvin cycle

3-PGA

NADPH + H$^+$

NADP$^+$

NADPH + H$^+$

Sucrose

Sucrose

(c)

(b)

10 μm

Copyright © The Benjamin/Cummings Publishing Company, Inc.

DNA as the Genetic Material of Bacteriophage T2

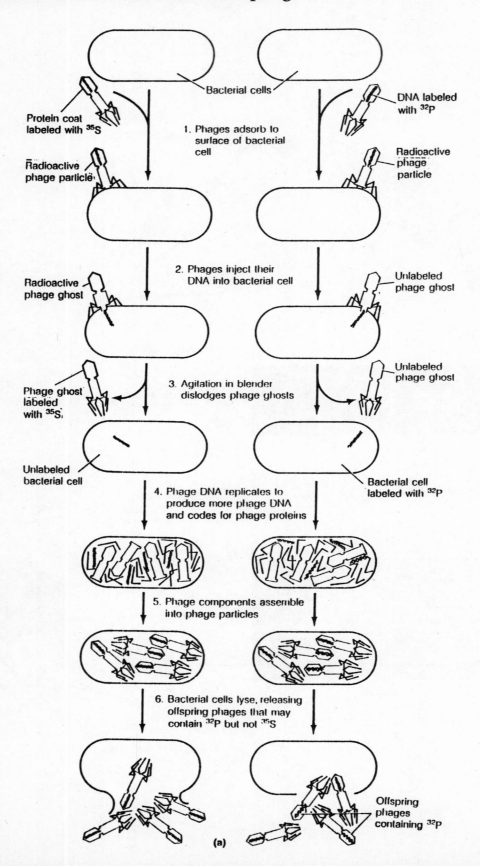

Bacterial cells

Protein coat labeled with ^{35}S

DNA labeled with ^{32}P

Radioactive phage particle

Radioactive phage particle

1. Phages adsorb to surface of bacterial cell

Radioactive phage ghost

Unlabeled phage ghost

2. Phages inject their DNA into bacterial cell

Phage ghost labeled with ^{35}S

Unlabeled phage ghost

3. Agitation in blender dislodges phage ghosts

Unlabeled bacterial cell

Bacterial cell labeled with ^{32}P

4. Phage DNA replicates to produce more phage DNA and codes for phage proteins

5. Phage components assemble into phage particles

6. Bacterial cells lyse, releasing offspring phages that may contain ^{32}P but not ^{35}S

Offspring phages containing ^{32}P

(a)

Copyright © The Benjamin/Cummings Publishing Company, Inc.

Thermal Denaturation Profile for DNA

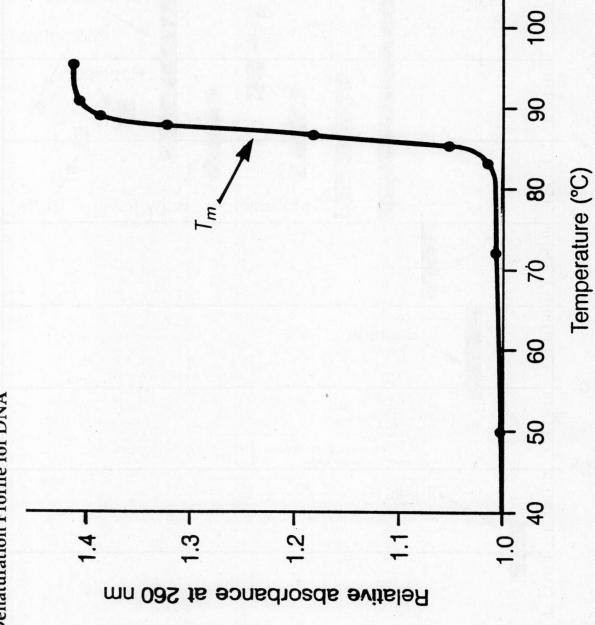

Copyright © The Benjamin/Cummings Publishing Company, Inc.

The Relationship Between Genome Size and Complexity of the Organism

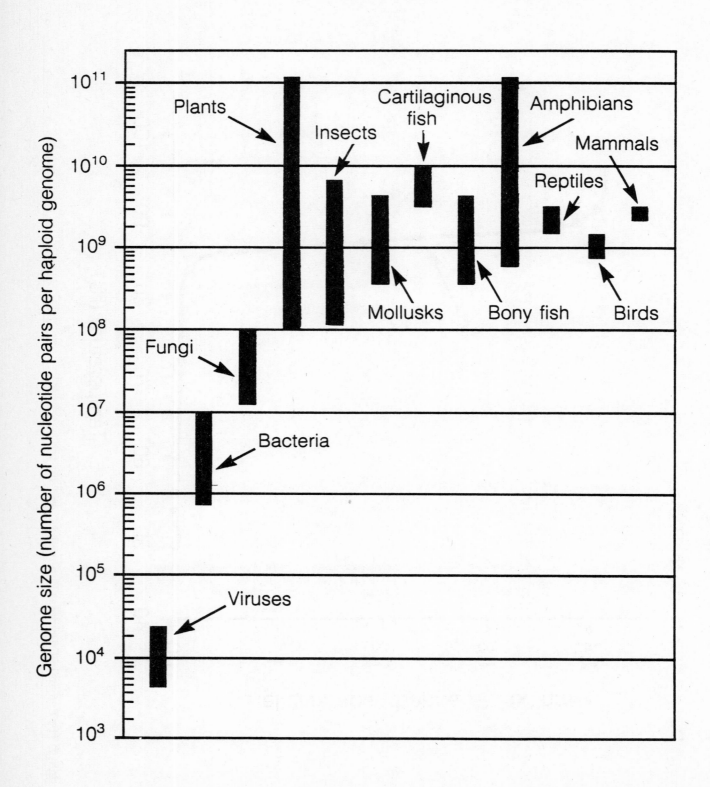

Copyright © The Benjamin/Cummings Publishing Company, Inc.

Preparation of DNA for Sequencing by the Chemical Method

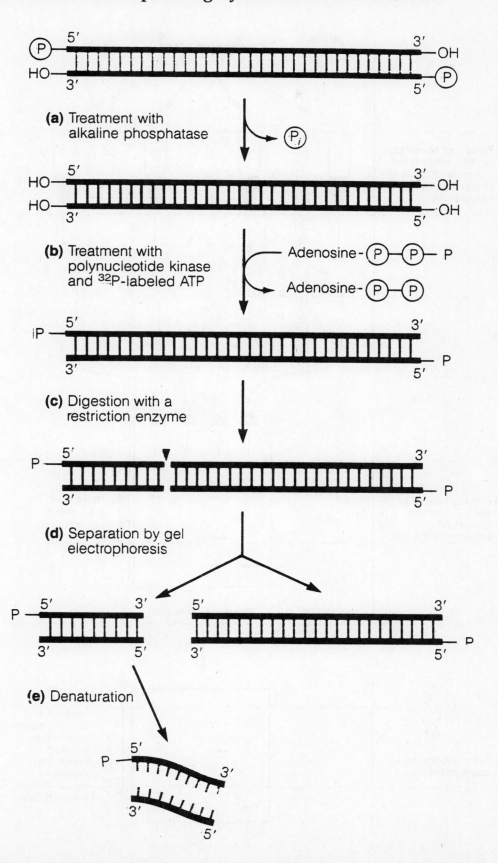

(a) Treatment with alkaline phosphatase

(b) Treatment with polynucleotide kinase and ³²P-labeled ATP

(c) Digestion with a restriction enzyme

(d) Separation by gel electrophoresis

(e) Denaturation

Copyright © The Benjamin/Cummings Publishing Company, Inc.

DNA Sequencing

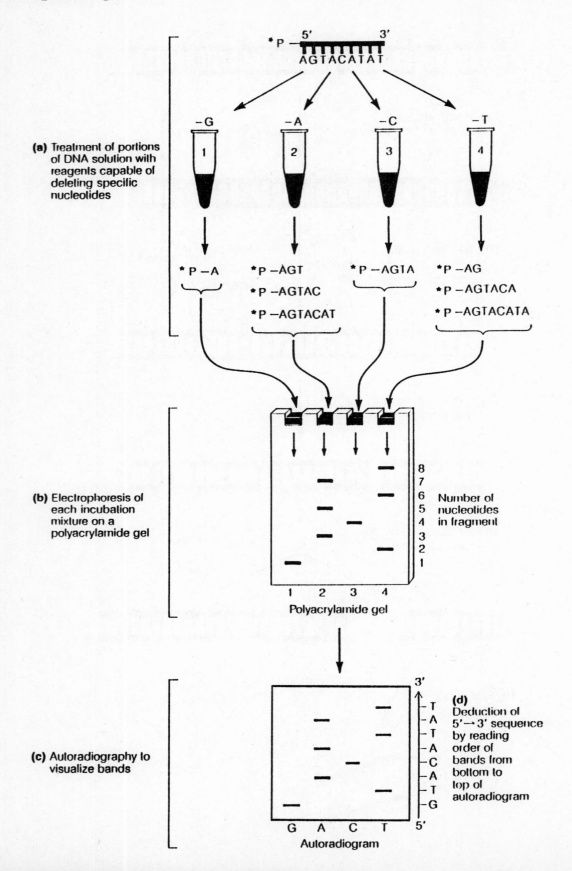

(a) Treatment of portions of DNA solution with reagents capable of deleting specific nucleotides

(b) Electrophoresis of each incubation mixture on a polyacrylamide gel

(c) Autoradiography to visualize bands

(d) Deduction of 5'→3' sequence by reading order of bands from bottom to top of autoradiogram

Number of nucleotides in fragment

Polyacrylamide gel

Autoradiogram

Copyright © The Benjamin/Cummings Publishing Company, Inc.

Nucleosome Structure

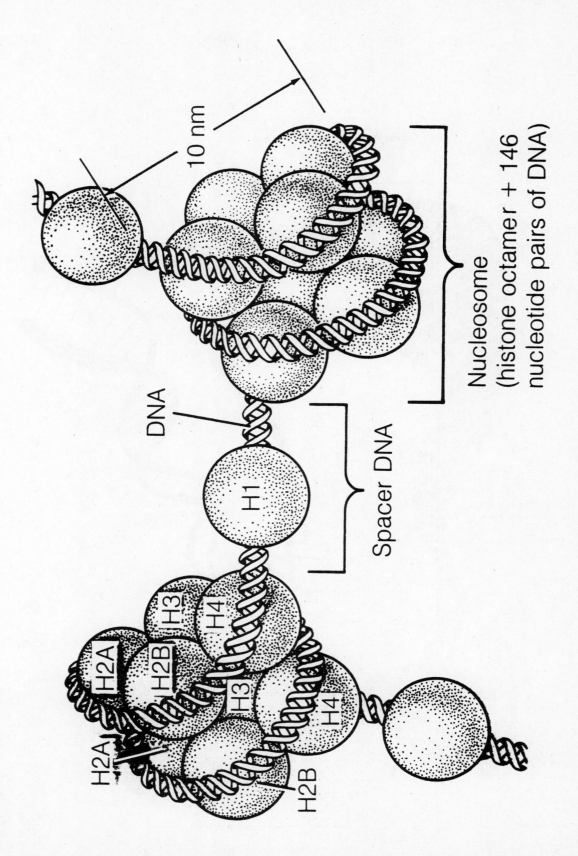

10 nm

DNA

H1

Spacer DNA

Nucleosome
(histone octamer + 146
nucleotide pairs of DNA)

H2A

H2B

H3

H4

H3

H4

H2A

H2B

Copyright © The Benjamin/Cummings Publishing Company, Inc.

Packaging of Nucleosomes into Chromosomes

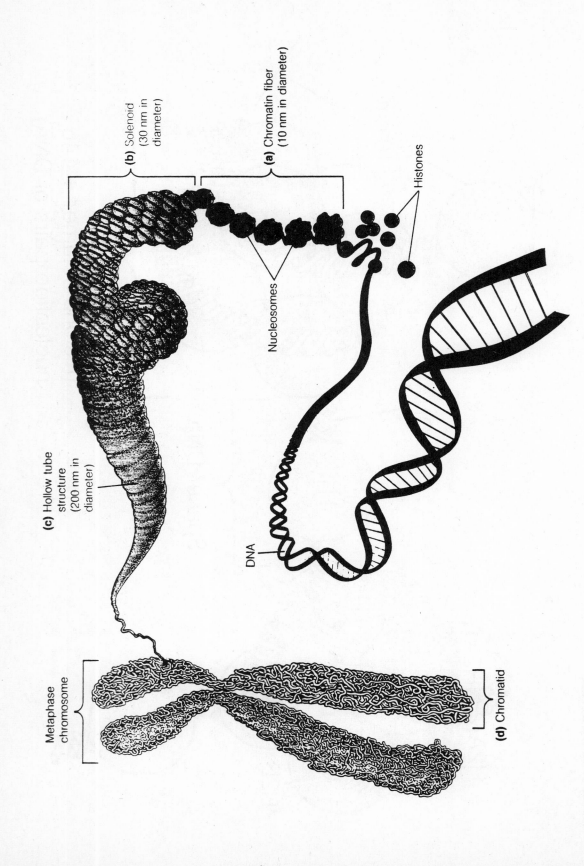

(b) Solenoid (30 nm in diameter)

(a) Chromatin fiber (10 nm in diameter)

Histones

Nucleosomes

(c) Hollow tube structure (200 nm in diameter)

DNA

Metaphase chromosome

(d) Chromatid

Copyright © The Benjamin/Cummings Publishing Company, Inc.

The Structure of the Nucleus

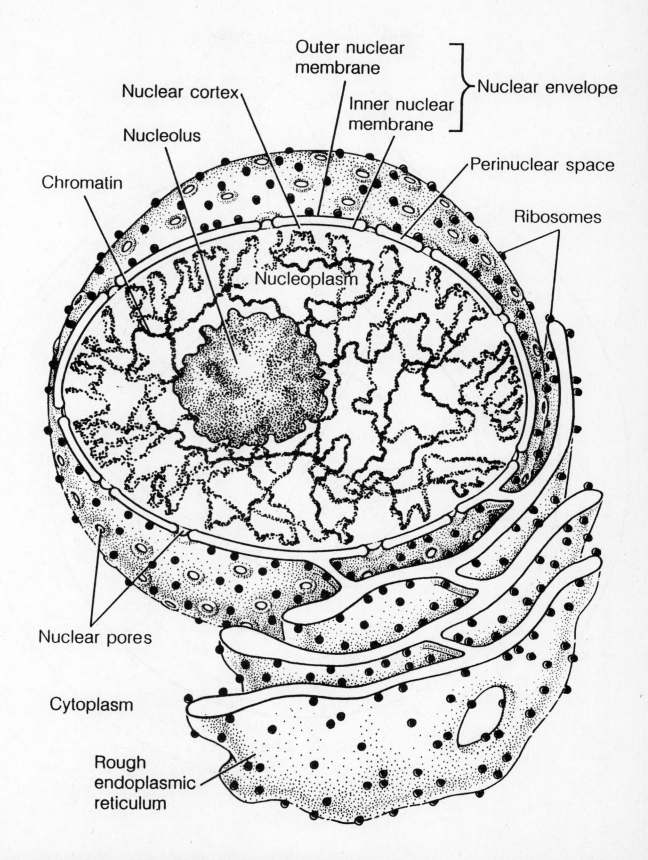

Nuclear cortex

Nucleolus

Chromatin

Outer nuclear membrane

Inner nuclear membrane

Nuclear envelope

Perinuclear space

Ribosomes

Nucleoplasm

Nuclear pores

Cytoplasm

Rough endoplasmic reticulum

Copyright © The Benjamin/Cummings Publishing Company, Inc.

The Eukaryotic Cell Cycle

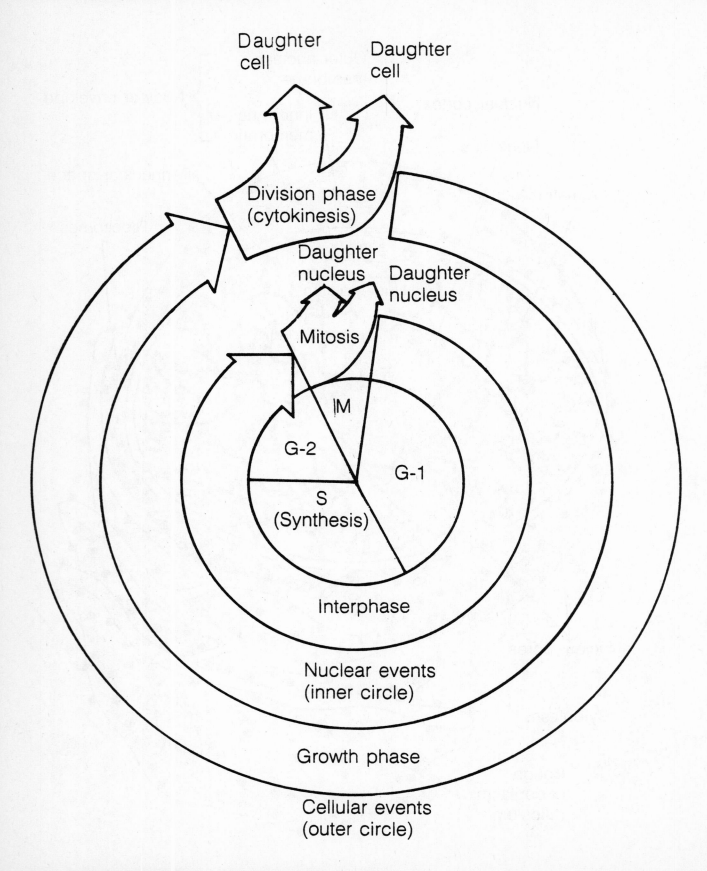

Daughter cell

Daughter cell

Division phase (cytokinesis)

Daughter nucleus

Daughter nucleus

Mitosis

M

G-2

G-1

S (Synthesis)

Interphase

Nuclear events (inner circle)

Growth phase

Cellular events (outer circle)

Copyright © The Benjamin/Cummings Publishing Company, Inc.

The Watson-Crick Model of DNA Replication

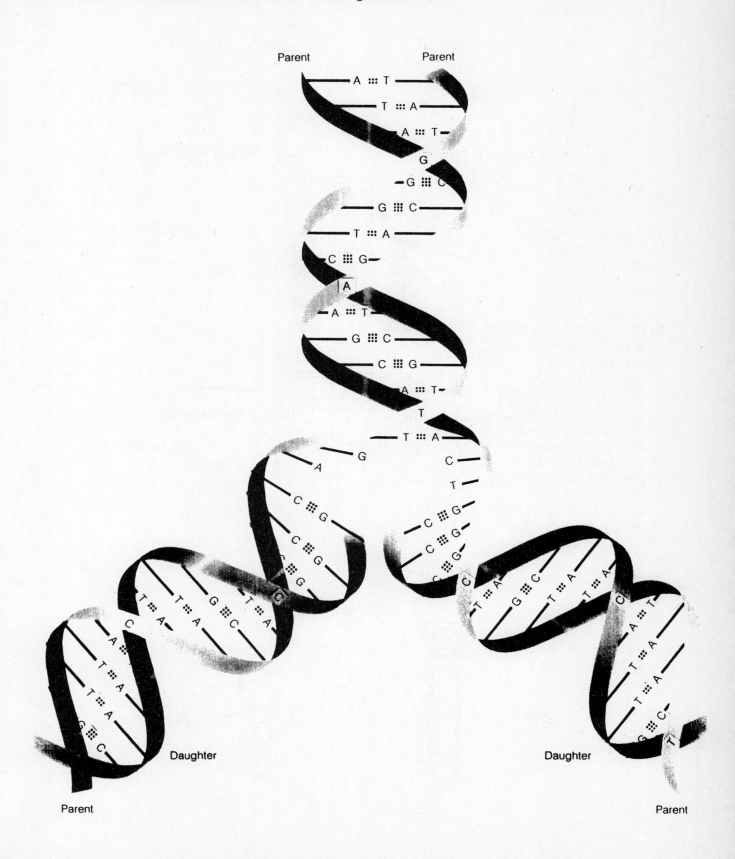

Copyright © The Benjamin/Cummings Publishing Company, Inc.

Equilibrium Density Centrifugation in DNA Analysis

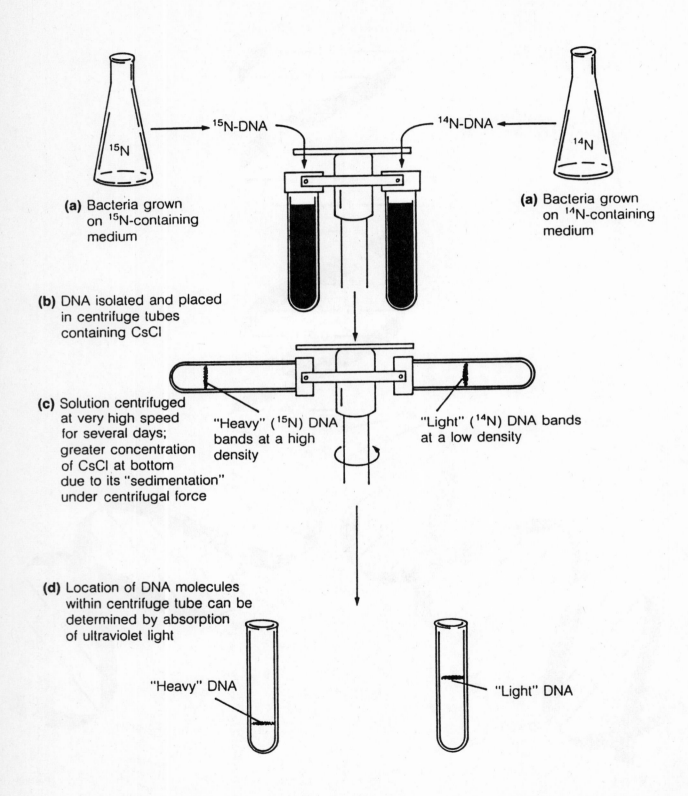

¹⁵N-DNA

¹⁴N-DNA

(a) Bacteria grown on ¹⁵N-containing medium

(a) Bacteria grown on ¹⁴N-containing medium

(b) DNA isolated and placed in centrifuge tubes containing CsCl

(c) Solution centrifuged at very high speed for several days; greater concentration of CsCl at bottom due to its "sedimentation" under centrifugal force

"Heavy" (¹⁵N) DNA bands at a high density

"Light" (¹⁴N) DNA bands at a low density

(d) Location of DNA molecules within centrifuge tube can be determined by absorption of ultraviolet light

"Heavy" DNA

"Light" DNA

Copyright © The Benjamin/Cummings Publishing Company, Inc.

Direction of DNA Synthesis at a Replication Fork

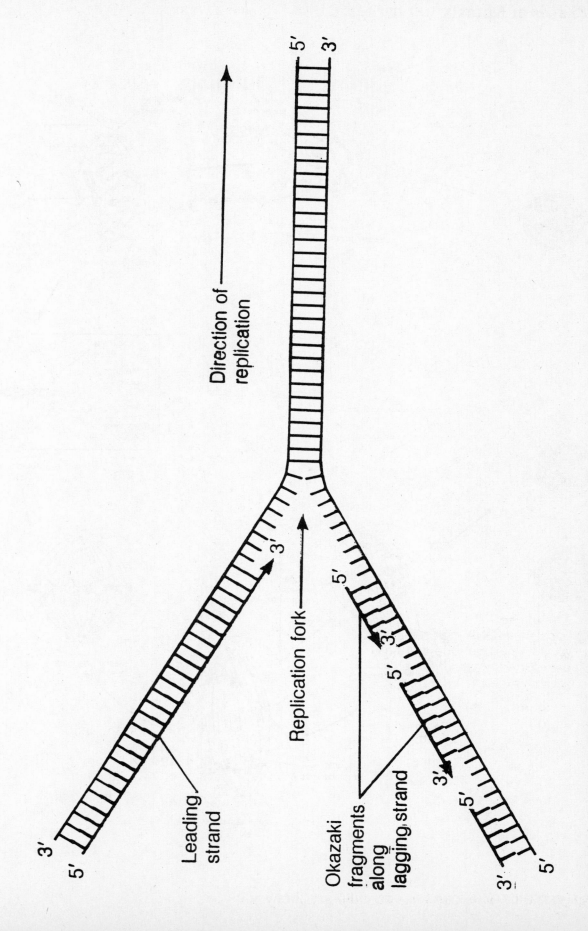

Direction of replication

Replication fork

Leading strand

Okazaki fragments along lagging strand

Copyright © The Benjamin/Cummings Publishing Company, Inc.

The Phases of Mitosis

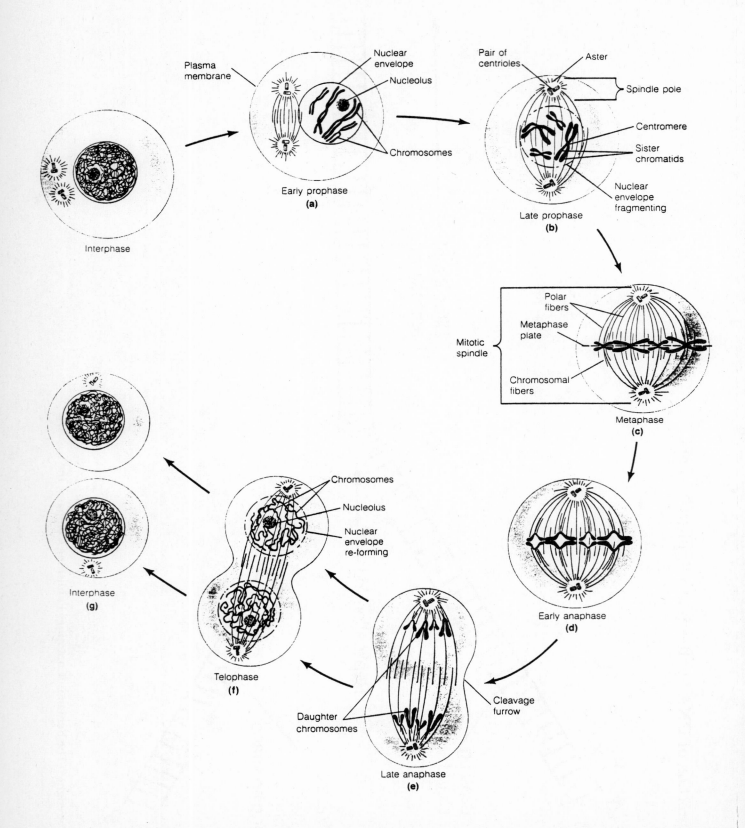

Plasma membrane

Nuclear envelope

Nucleolus

Chromosomes

Early prophase
(a)

Pair of centrioles

Aster

Spindle pole

Centromere

Sister chromatids

Nuclear envelope fragmenting

Late prophase
(b)

Interphase

Polar fibers

Metaphase plate

Mitotic spindle

Chromosomal fibers

Metaphase
(c)

Chromosomes

Nucleolus

Nuclear envelope re-forming

Interphase
(g)

Telophase
(f)

Daughter chromosomes

Early anaphase
(d)

Cleavage furrow

Late anaphase
(e)

Copyright © The Benjamin/Cummings Publishing Company, Inc.

Overview of Meiosis

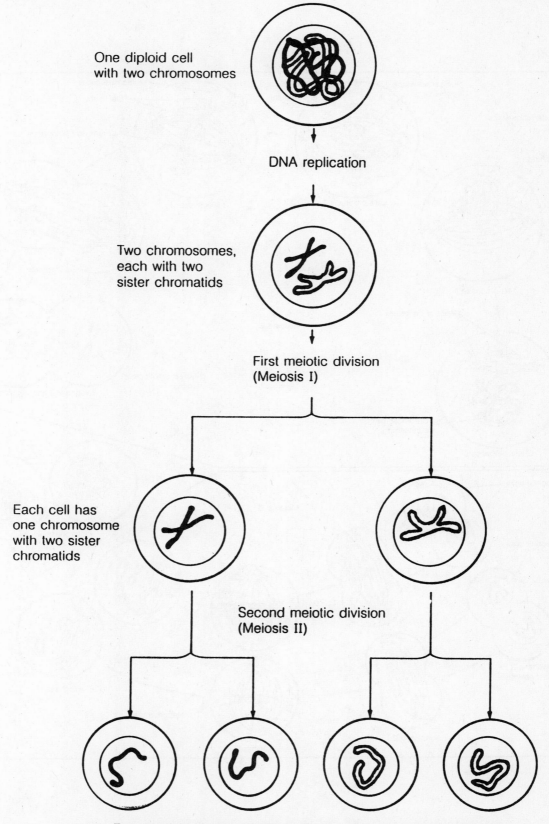

One diploid cell
with two chromosomes

DNA replication

Two chromosomes,
each with two
sister chromatids

First meiotic division
(Meiosis I)

Each cell has
one chromosome
with two sister
chromatids

Second meiotic division
(Meiosis II)

Four haploid daughter cells with one chromosome in each cell

Copyright © The Benjamin/Cummings Publishing Company, Inc.

Meiosis

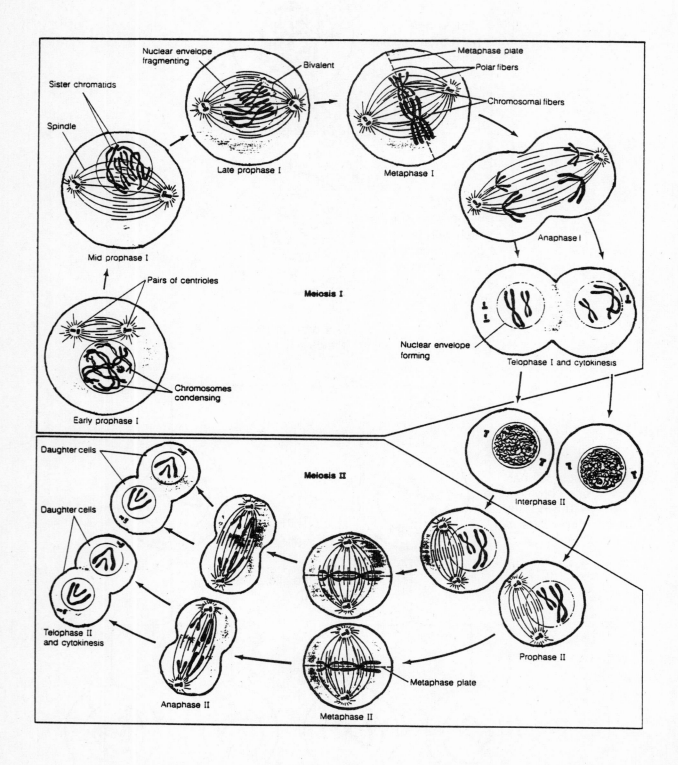

Copyright © The Benjamin/Cummings Publishing Company, Inc.

Comparison of Meiosis and Mitosis

Meiosis

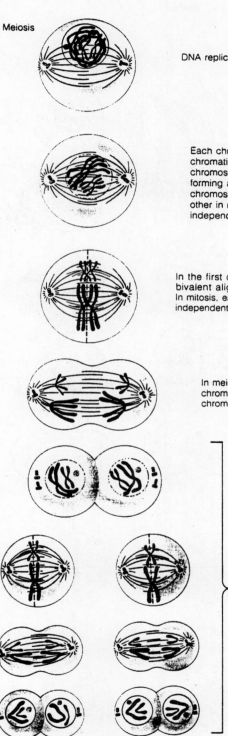

Mitosis

DNA replicates. Meiosis or mitosis begins.

Each chromosome now has two chromatids. In meiosis, homologous chromosomes attach to each other, forming a bivalent. Homologous chromosomes do not attach to each other in mitosis, and thus act independently.

In the first division of meiosis, each bivalent aligns at the metaphase plate. In mitosis, each chromosome aligns independently at the metaphase plate.

In meiosis I, chromosomes (not chromatids) separate. In mitosis, chromatids separate.

In the second division of meiosis, sister chromatids separate.

Result of mitosis: two cells, each with the same number of chromosomes as the original cell.

Result of meiosis: four haploid cells, each with half as many chromosomes as the original cell.

Copyright © The Benjamin/Cummings Publishing Company, Inc.

Meiosis and Gamete Formation

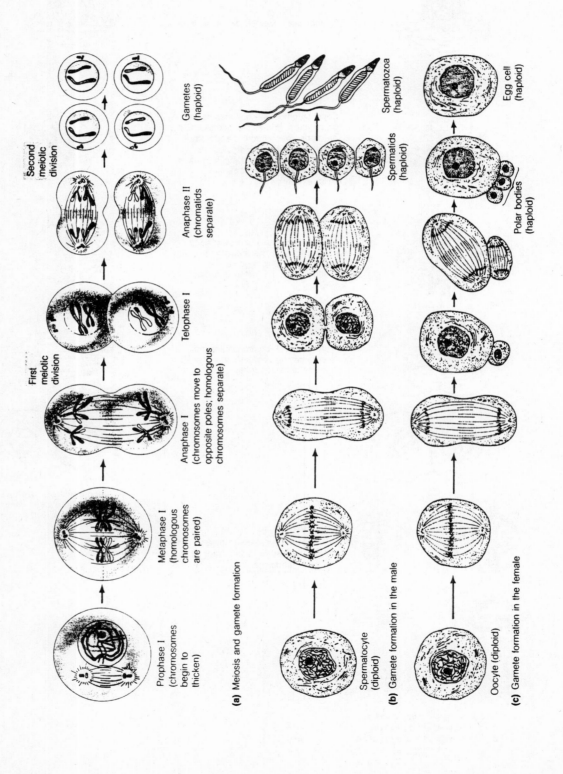

First meiotic division

Second meiotic division

Prophase I (chromosomes begin to thicken)

Metaphase I (homologous chromosomes are paired)

Anaphase I (chromosomes move to opposite poles; homologous chromosomes separate)

Telophase I

Anaphase II (chromatids separate)

Gametes (haploid)

(a) Meiosis and gamete formation

Spermatocyte (diploid)

Spermatids (haploid)

Spermatozoa (haploid)

(b) Gamete formation in the male

Oocyte (diploid)

Polar bodies (haploid)

Egg cell (haploid)

(c) Gamete formation in the female

Copyright © The Benjamin/Cummings Publishing Company, Inc.

The Law of Segregation

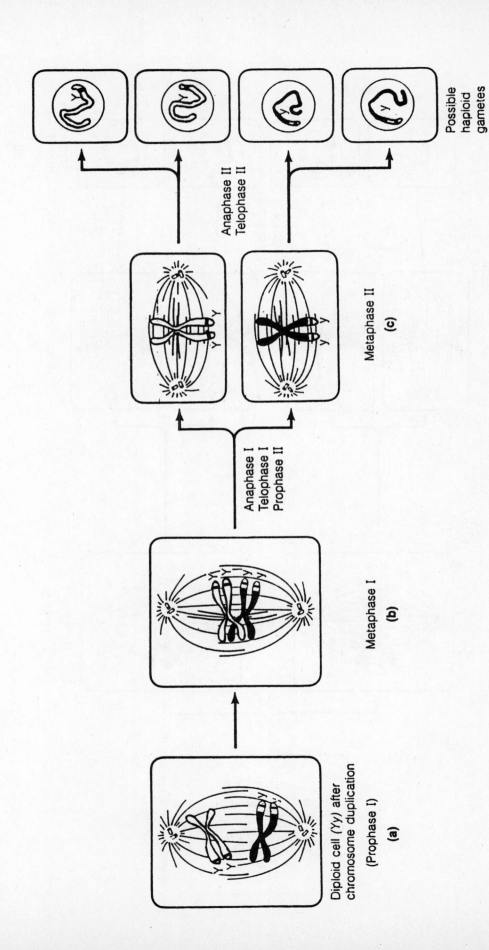

Diploid cell (Yy) after
chromosome duplication
(Prophase I)
(a)

Metaphase I
(b)

Anaphase I
Telophase I
Prophase II

Metaphase II
(c)

Anaphase II
Telophase II

Possible
haploid
gametes
(d)

Copyright © The Benjamin/Cummings Publishing Company, Inc.

The Law of Independent Assortment

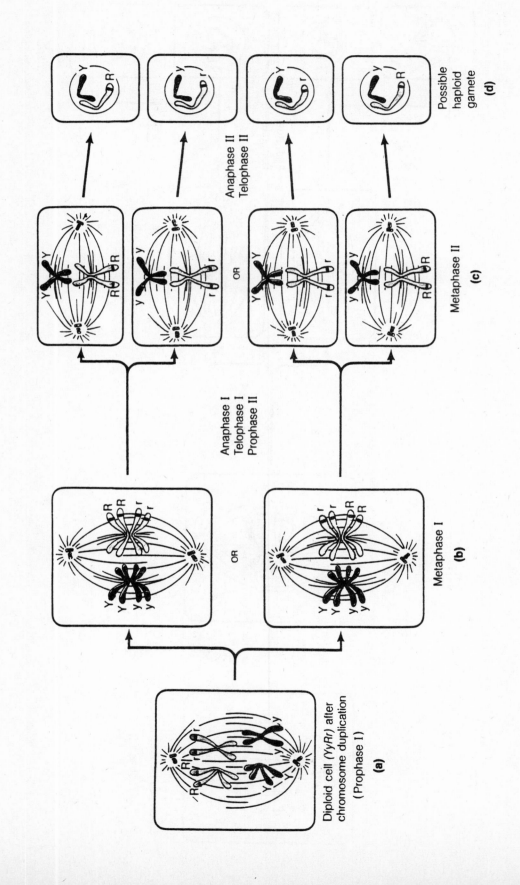

Diploid cell (YyRr) after chromosome duplication (Prophase 1)

(a)

Metaphase I

(b)

OR

Anaphase I
Telophase I
Prophase II

Metaphase II

(c)

OR

Anaphase II
Telophase II

Possible haploid gamete

(d)

Copyright © The Benjamin/Cummings Publishing Company, Inc.

Transformation and Transduction in Bacterial Cells

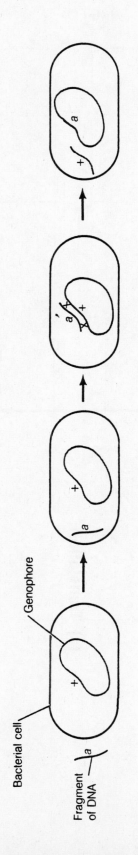

(a) Transformation of a bacterial cell by exogenous DNA

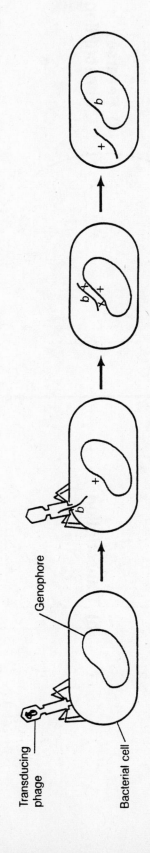

(b) Transduction of a bacterial cell by a transducing phage

Copyright © The Benjamin/Cummings Publishing Company, Inc.

The Central Dogma of Molecular Biology

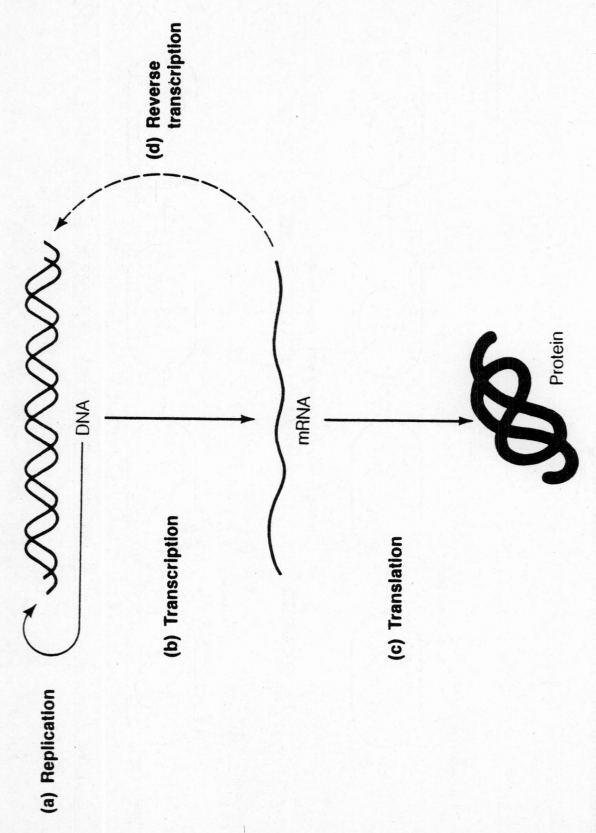

(a) Replication

DNA

(b) Transcription

mRNA

(c) Translation

Protein

(d) Reverse transcription

Copyright © The Benjamin/Cummings Publishing Company, Inc.

RNAs as Intermediates in Information flow

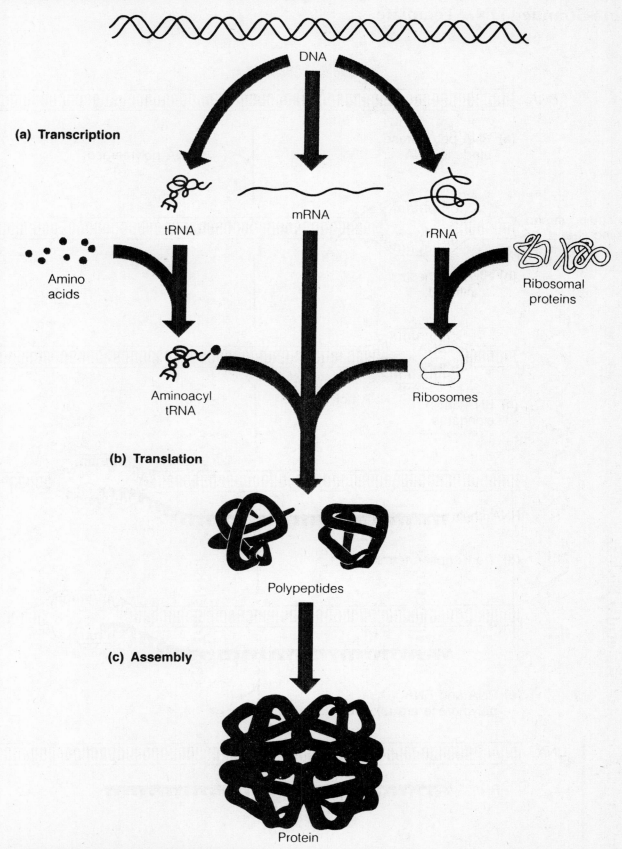

DNA

(a) Transcription

tRNA

mRNA

rRNA

Amino
acids

Ribosomal
proteins

Aminoacyl
tRNA

Ribosomes

(b) Translation

Polypeptides

(c) Assembly

Protein

Copyright © The Benjamin/Cummings Publishing Company, Inc.

The Transcription of Single-Stranded RNA from a
Double-Stranded DNA Template

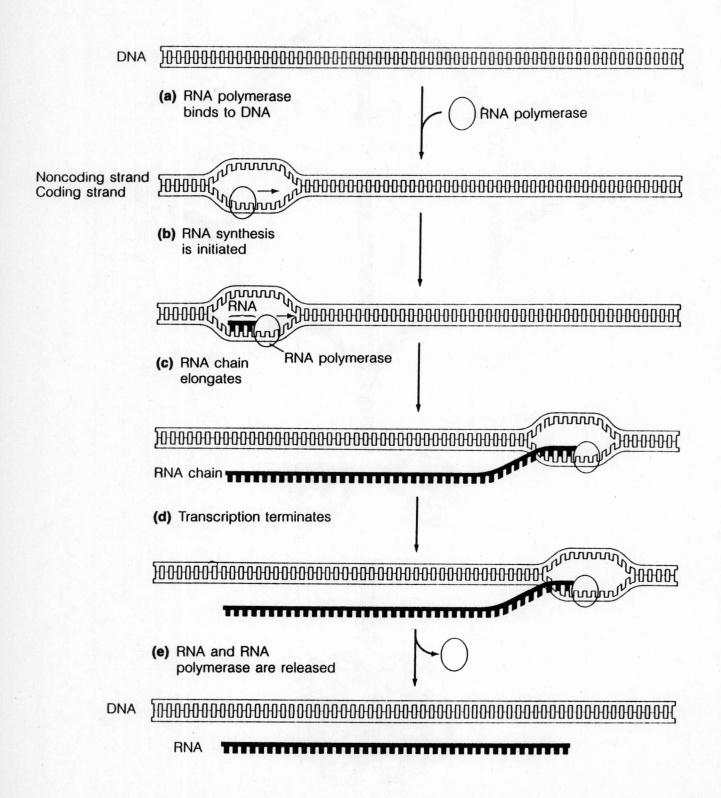

DNA

(a) RNA polymerase
binds to DNA

RNA polymerase

Noncoding strand
Coding strand

(b) RNA synthesis
is initiated

RNA

(c) RNA chain
elongates

RNA polymerase

RNA chain

(d) Transcription terminates

(e) RNA and RNA
polymerase are released

DNA

RNA

Copyright © The Benjamin/Cummings Publishing Company, Inc.

The Genetic Code

Second Position

First Position		U	C	A	G		Third Position
U		UUU ⎤ Phe UUC ⎦ UUA ⎤ Leu UUG ⎦	UCU ⎤ UCC UCA Ser UCG ⎦	UAU ⎤ Tyr UAC ⎦ UAA Stop UAG Stop	UGU ⎤ Cys UGC ⎦ UGA Stop UGG Trp		U C A G
C		CUU ⎤ CUC CUA Leu CUG ⎦	CCU ⎤ CCC CCA Pro CCG ⎦	CAU ⎤ His CAC ⎦ CAA ⎤ Gln CAG ⎦	CGU ⎤ CGC CGA Arg CGG ⎦		U C A G
A		AUU ⎤ AUC Ile AUA ⎦ AUG Met	ACU ⎤ ACC ACA Thr ACG ⎦	AAU ⎤ Asn AAC ⎦ AAA ⎤ Lys AAG ⎦	AGU ⎤ Ser AGC ⎦ AGA ⎤ Arg AGG ⎦		U C A G
G		GUU ⎤ GUC GUA Val GUG ⎦	GCU ⎤ GCC GCA Ala GCG ⎦	GAU ⎤ Asp GAC ⎦ GAA ⎤ Glu GAG ⎦	GGU ⎤ GGC GGA Gly GGG ⎦		U C A G

Copyright © The Benjamin/Cummings Publishing Company, Inc.

Transcription in Bacterial Cells

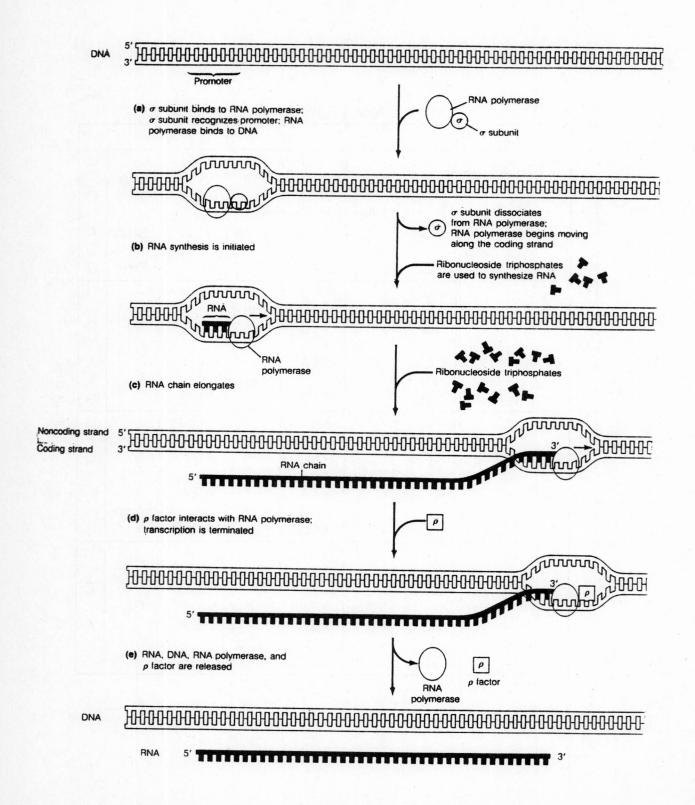

Copyright © The Benjamin/Cummings Publishing Company, Inc.

Retrovirus Infection and Reverse Transcription

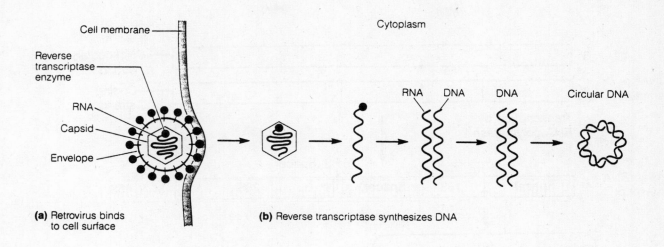

Cell membrane

Reverse transcriptase enzyme

RNA

Capsid

Envelope

(a) Retrovirus binds to cell surface

Cytoplasm

RNA   DNA   DNA   Circular DNA

(b) Reverse transcriptase synthesizes DNA

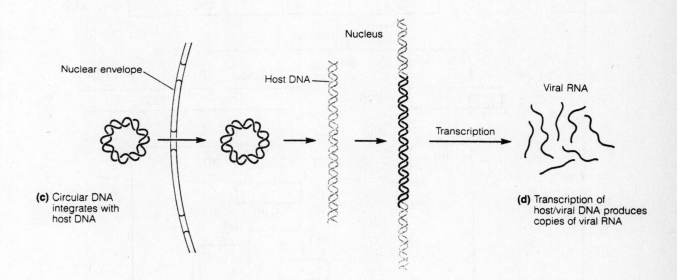

Nuclear envelope

Nucleus

Host DNA

DNA

Viral RNA

Transcription

(c) Circular DNA integrates with host DNA

(d) Transcription of host/viral DNA produces copies of viral RNA

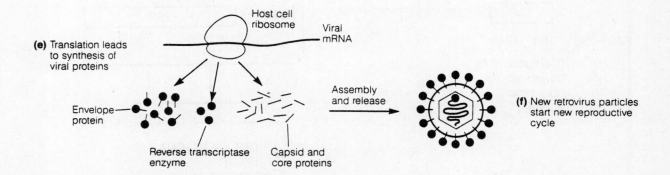

Host cell ribosome

Viral mRNA

(e) Translation leads to synthesis of viral proteins

Envelope protein

Reverse transcriptase enzyme

Capsid and core proteins

Assembly and release

(f) New retrovirus particles start new reproductive cycle

Copyright © The Benjamin/Cummings Publishing Company, Inc.

Processing of Ribosomal RNA

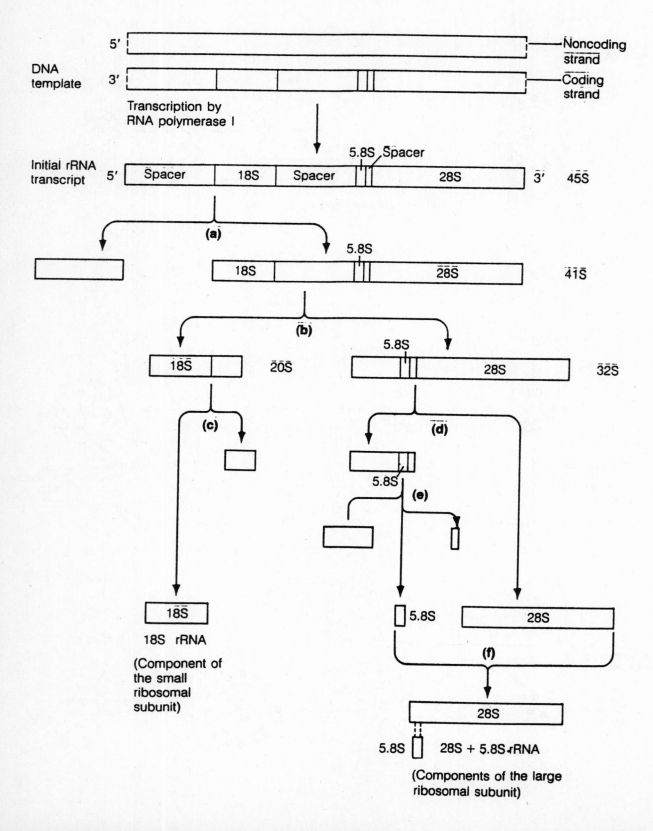

Copyright © The Benjamin/Cummings Publishing Company, Inc.

Processing and Structure of Transfer RNA

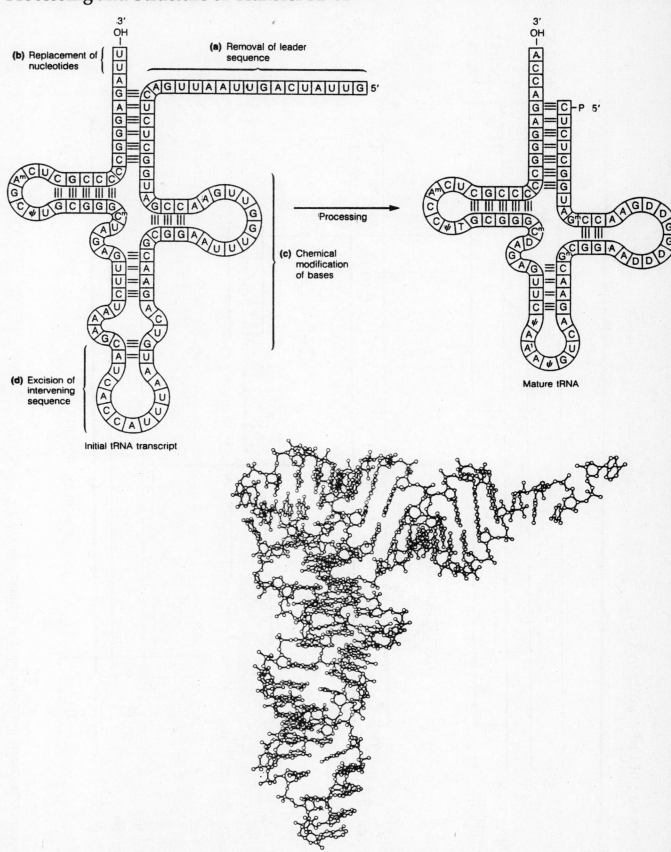

(b) Replacement of nucleotides

(a) Removal of leader sequence

Processing

(c) Chemical modification of bases

(d) Excision of intervening sequence

Initial tRNA transcript

Mature tRNA

Copyright © The Benjamin/Cummings Publishing Company, Inc.

Processing of Eukaryotic mRNA

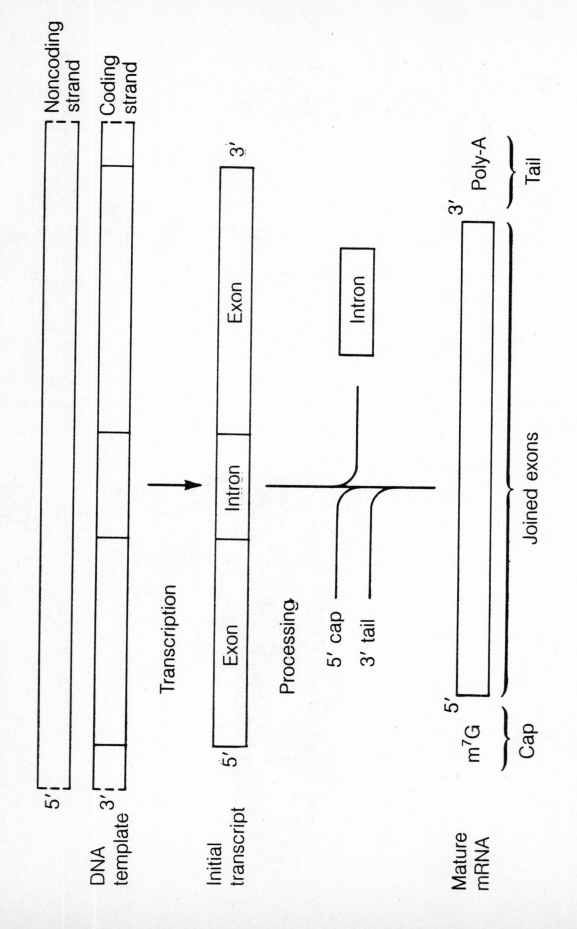

Copyright © The Benjamin/Cummings Publishing Company, Inc.

Sequence, Structure, and Aminoacylation of tRNA

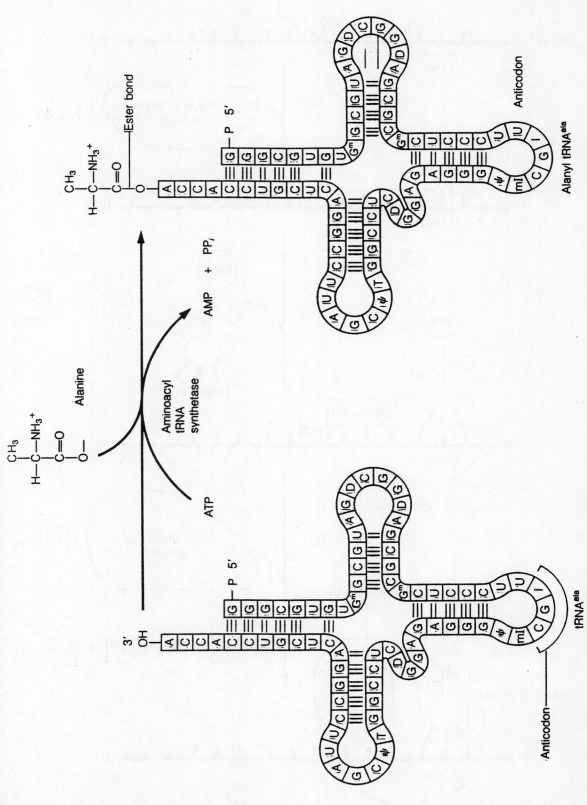

Copyright © The Benjamin/Cummings Publishing Company, Inc.

The Initiation Complex as a Prerequisite for Protein Synthesis in Prokaryotes

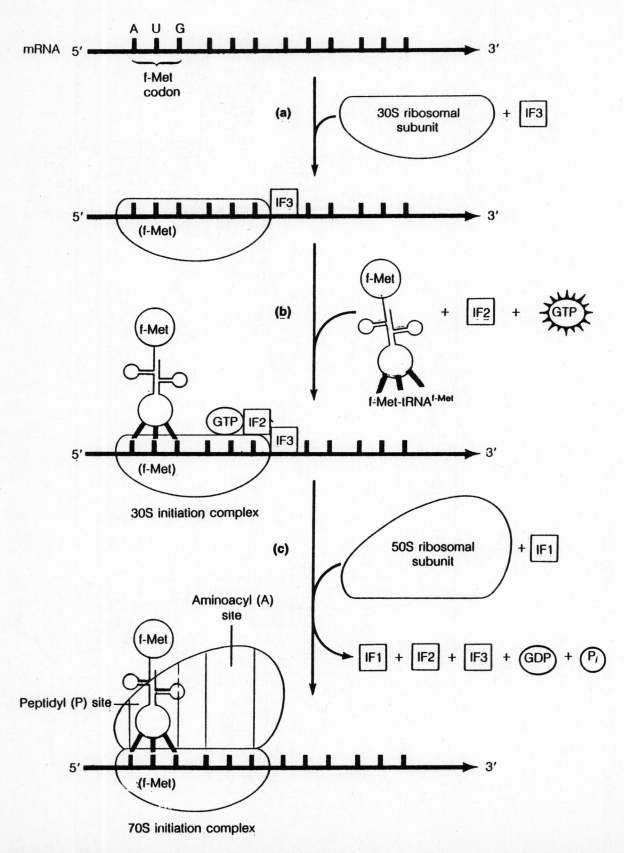

Copyright © The Benjamin/Cummings Publishing Company, Inc.

Polypeptide Chain Elongation

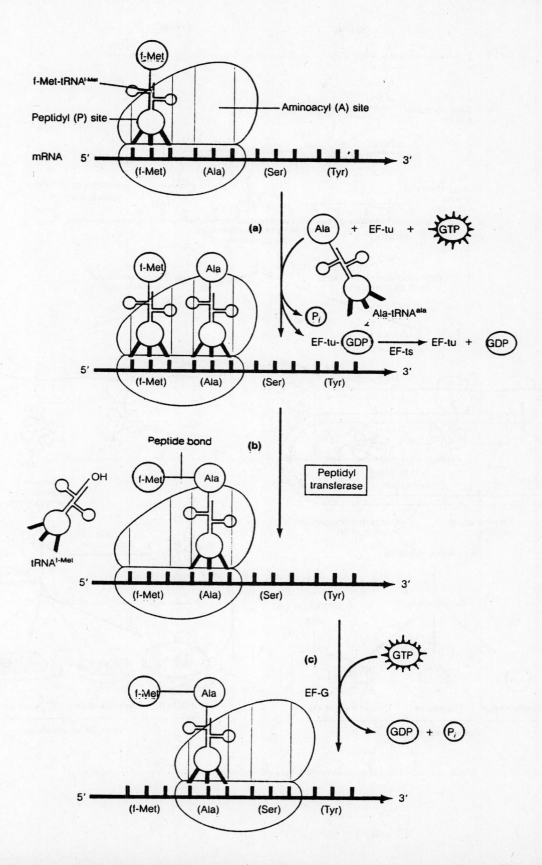

Copyright © The Benjamin/Cummings Publishing Company, Inc.

Intracellular Sorting of Proteins

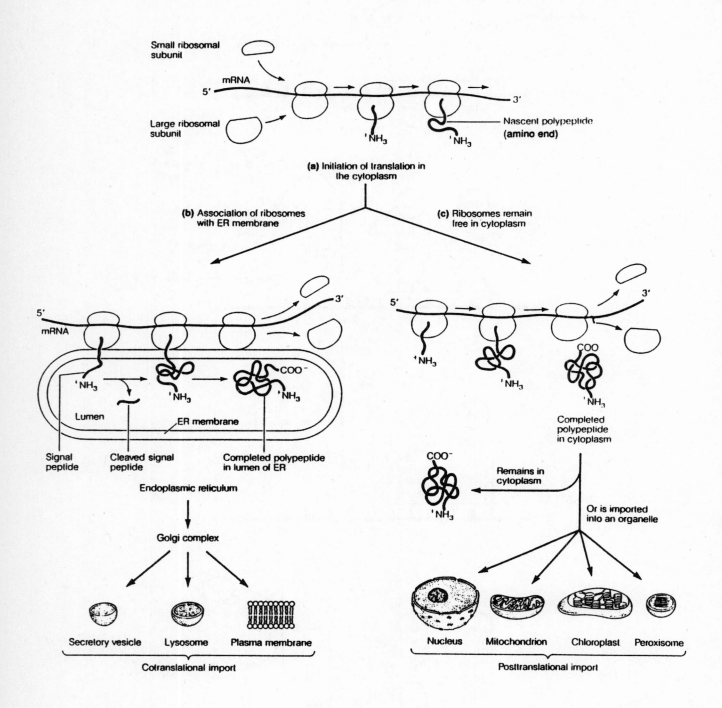

Small ribosomal subunit

mRNA

5′

Large ribosomal subunit

'NH₃

'NH₃

Nascent polypeptide (amino end)

3′

(a) Initiation of translation in the cytoplasm

(b) Association of ribosomes with ER membrane

(c) Ribosomes remain free in cytoplasm

5′

mRNA

3′

'NH₃

'NH₃

COO⁻

'NH₃

Lumen

ER membrane

Signal peptide

Cleaved signal peptide

Completed polypeptide in lumen of ER

Endoplasmic reticulum

Golgi complex

Secretory vesicle   Lysosome   Plasma membrane

Cotranslational import

5′

'NH₃

'NH₃

COO

'NH₃

3′

Completed polypeptide in cytoplasm

COO⁻

'NH₃

Remains in cytoplasm

Or is imported into an organelle

Nucleus   Mitochondrion   Chloroplast   Peroxisome

Posttranslational import

Copyright © The Benjamin/Cummings Publishing Company, Inc.

A Typical Anabolic Pathway

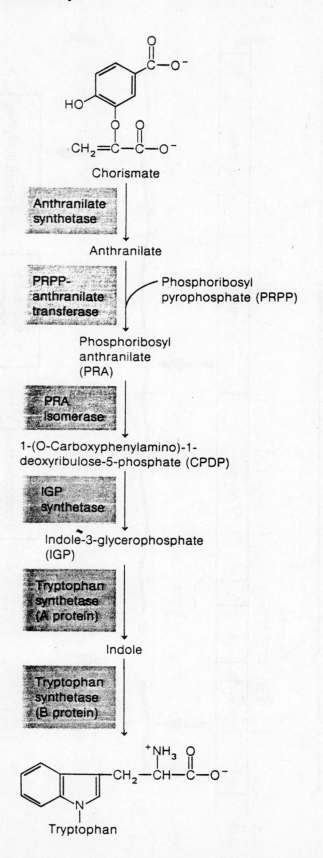

Copyright © The Benjamin/Cummings Publishing Company, Inc.

The Lactose (*lac*) Operon of *E. coli*

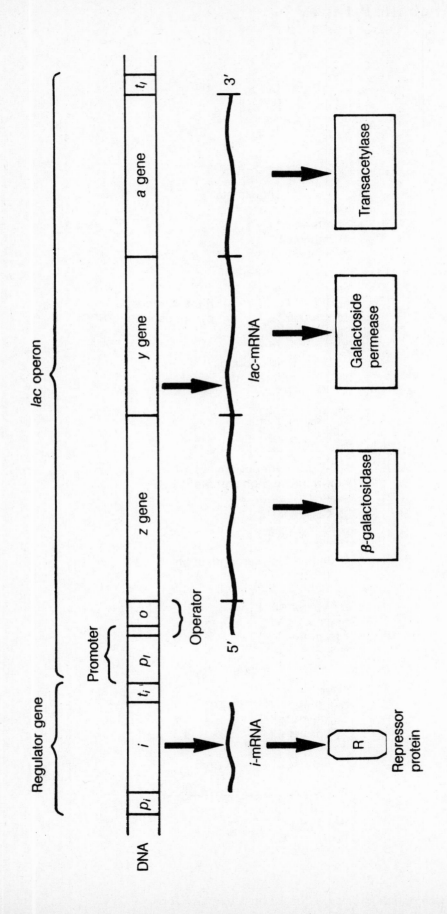

Copyright © The Benjamin/Cummings Publishing Company, Inc.

Allosteric Regulation of the Lactose Repressor

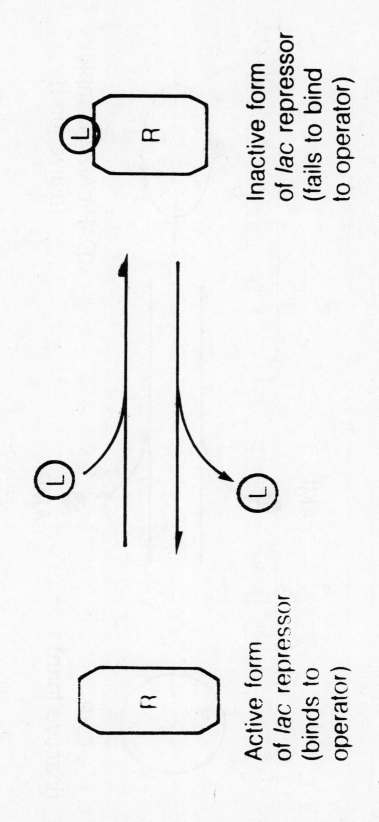

Active form
of *lac* repressor
(binds to
operator)

Inactive form
of *lac* repressor
(fails to bind
to operator)

Copyright © The Benjamin/Cummings Publishing Company, Inc.

Allosteric Regulation of the Catabolite Activator Protein by Cyclic AMP

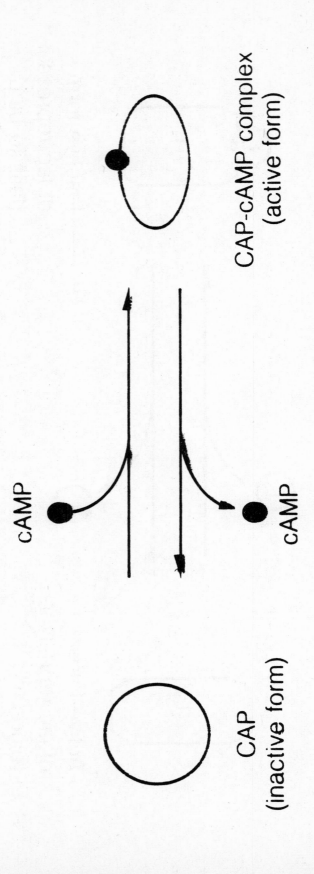

Copyright © The Benjamin/Cummings Publishing Company, Inc.

Differential Transcription Can Be Demonstrated by Nuclear Run-on Transcription Assays

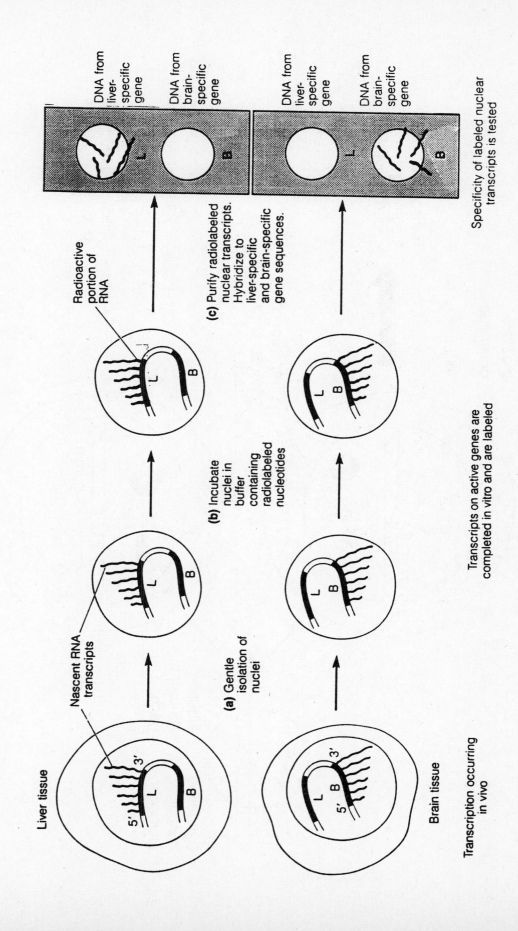

Copyright © The Benjamin/Cummings Publishing Company, Inc.

Combinatorial Model for Gene Expression

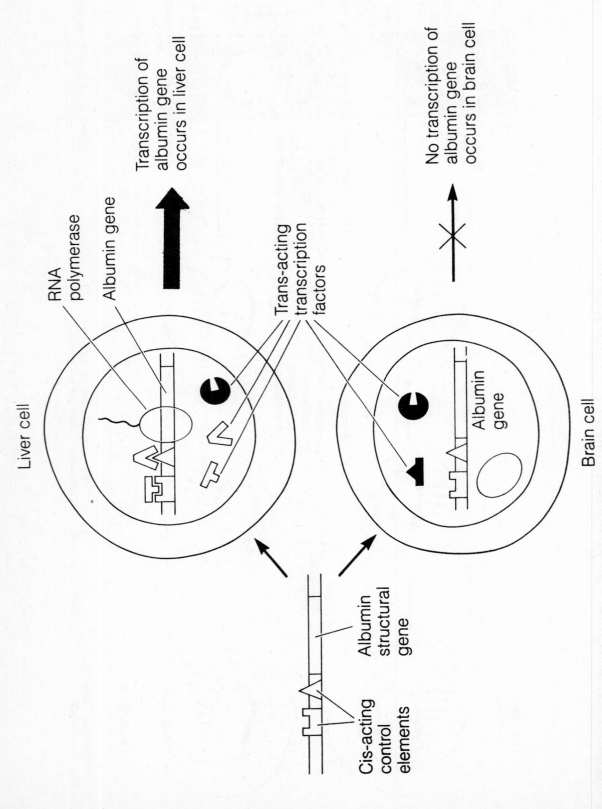

Liver cell

RNA polymerase

Albumin gene

Transcription of albumin gene occurs in liver cell

Trans-acting transcription factors

No transcription of albumin gene occurs in brain cell

Albumin gene

Brain cell

Albumin structural gene

Cis-acting control elements

Copyright © The Benjamin/Cummings Publishing Company, Inc.

Diagram of a Generalized Eukaryotic Upstream Promoter Region and Transcription Unit

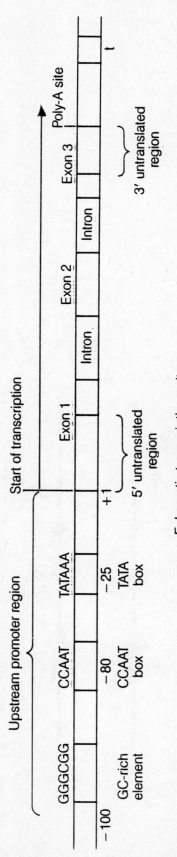

Copyright © The Benjamin/Cummings Publishing Company, Inc.

Properties of Euykaryotic Enhancer Elements

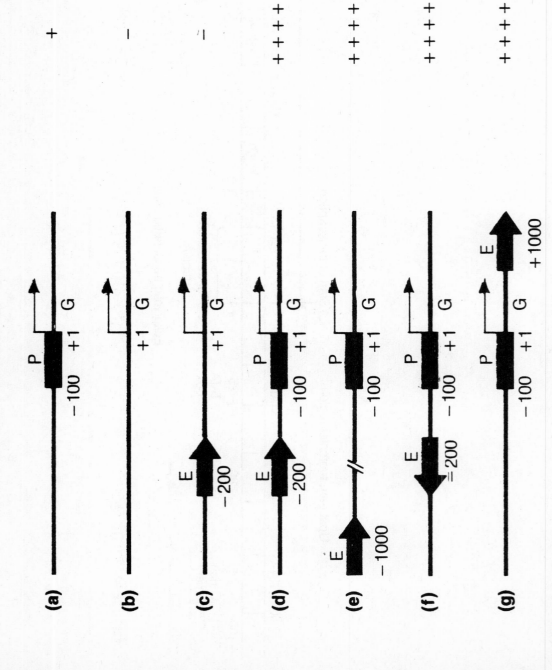

Copyright © The Benjamin/Cummings Publishing Company, Inc.

Common Structural Motifs in Transcription Factors

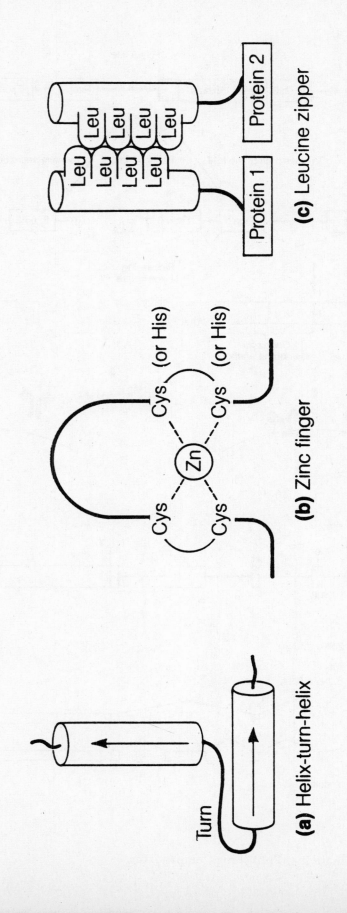

(a) Helix-turn-helix

(b) Zinc finger

(c) Leucine zipper

Copyright © The Benjamin/Cummings Publishing Company, Inc.

Alternative Splicing to Produce Variant Gene Products

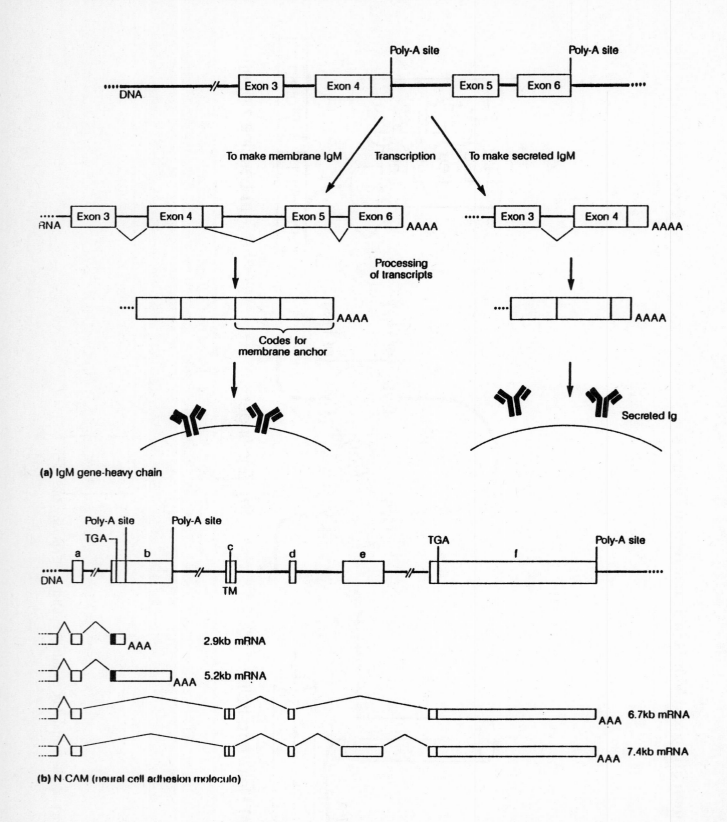

(a) IgM gene-heavy chain

(b) N-CAM (neural cell adhesion molecule)

Copyright © The Benjamin/Cummings Publishing Company, Inc.

Regulation of Translation by Hemin in Red Blood Cells

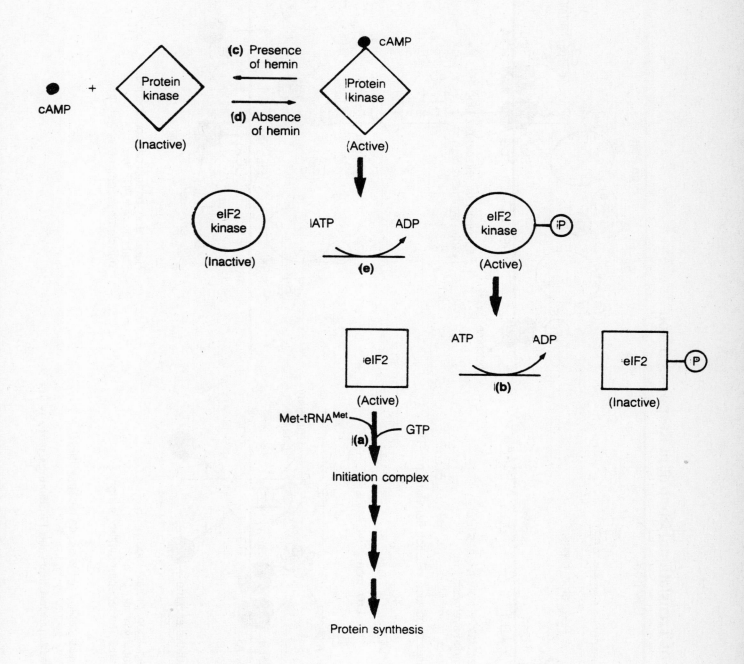

Copyright © The Benjamin/Cummings Publishing Company, Inc.

Two Levels of Translational Control in Response to Iron

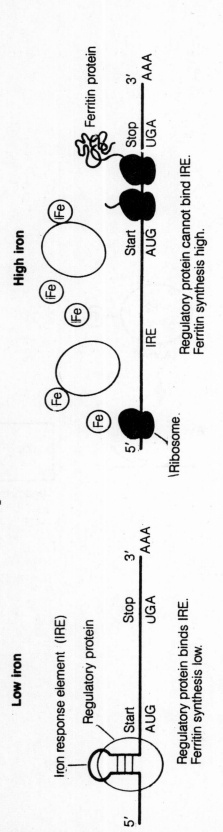

(a) Iron increases translation initiation from ferritin mRNA

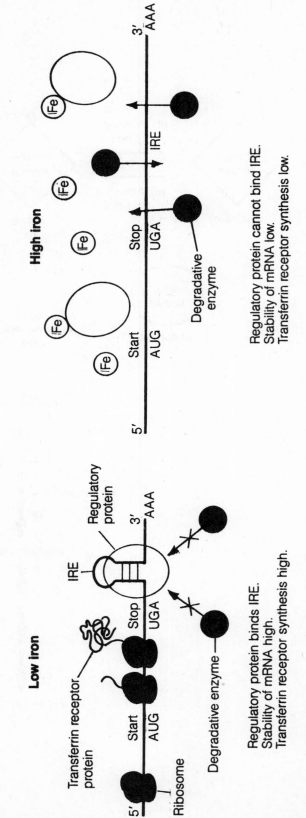

(b) Iron decreases stability of transferrin receptor mRNA

Copyright © The Benjamin/Cummings Publishing Company, Inc.

Microtubules

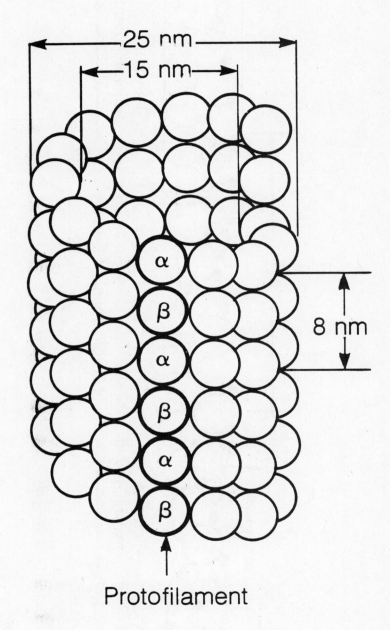

Protofilament

Copyright © The Benjamin/Cummings Publishing Company, Inc.

Assembly of Microfilaments

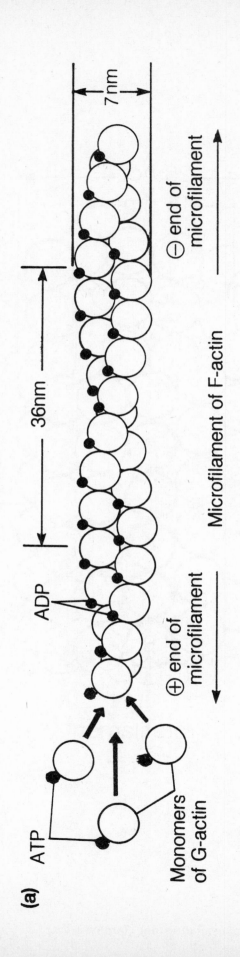

(a)

ATP

Monomers
of G-actin

⊕ end of
microfilament

ADP

7 nm

36nm

⊖ end of
microfilament

Microfilament of F-actin

Copyright © The Benjamin/Cummings Publishing Company, Inc.

Treadmilling of Actin Microfilaments

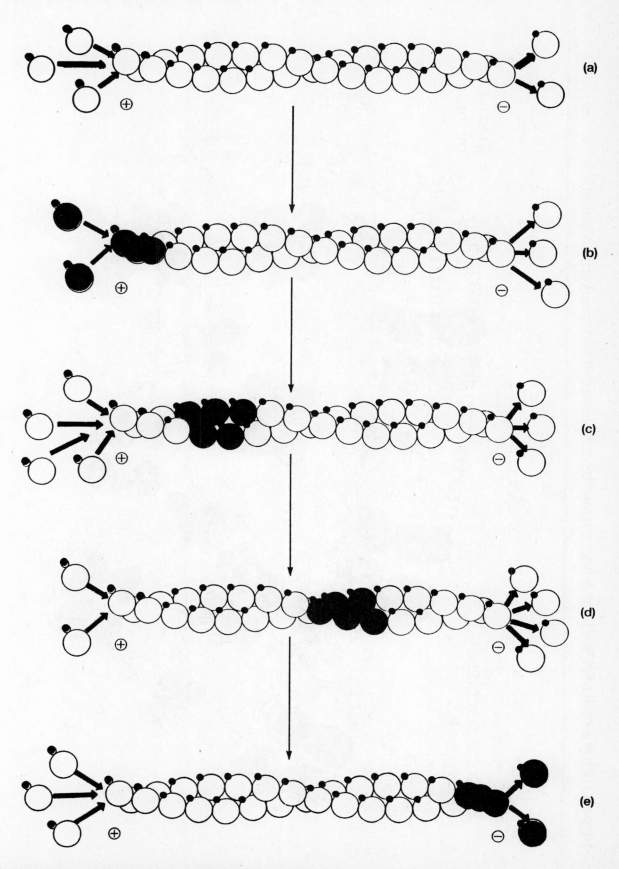

(a)

(b)

(c)

(d)

(e)

Copyright © The Benjamin/Cummings Publishing Company, Inc.

Support of the Erythrocyte Plasma Membrane by a Spectrin-Ankyrin-Actin Network

Plasma membrane

Short actin chains

Ankyrin

Transmembrane proteins

Spectrin filaments

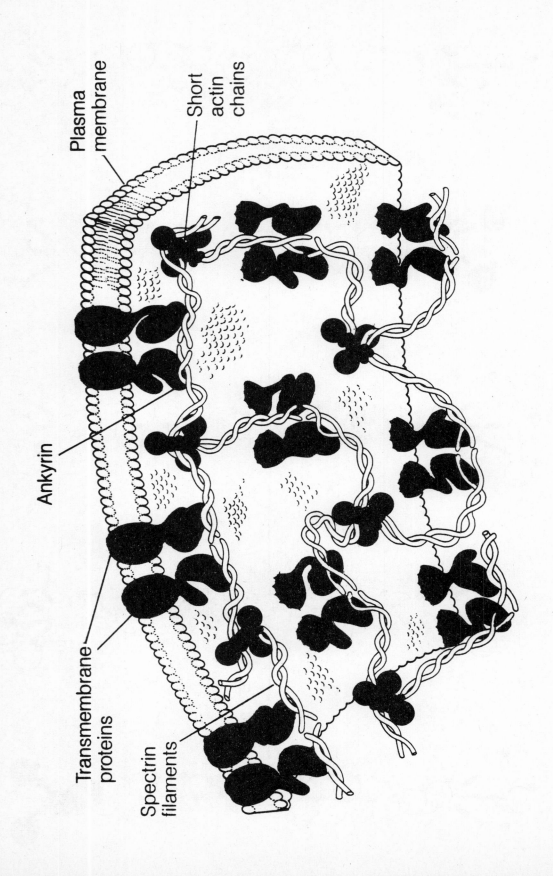

Copyright © The Benjamin/Cummings Publishing Company, Inc.

Model for Intermediate Filament Assembly

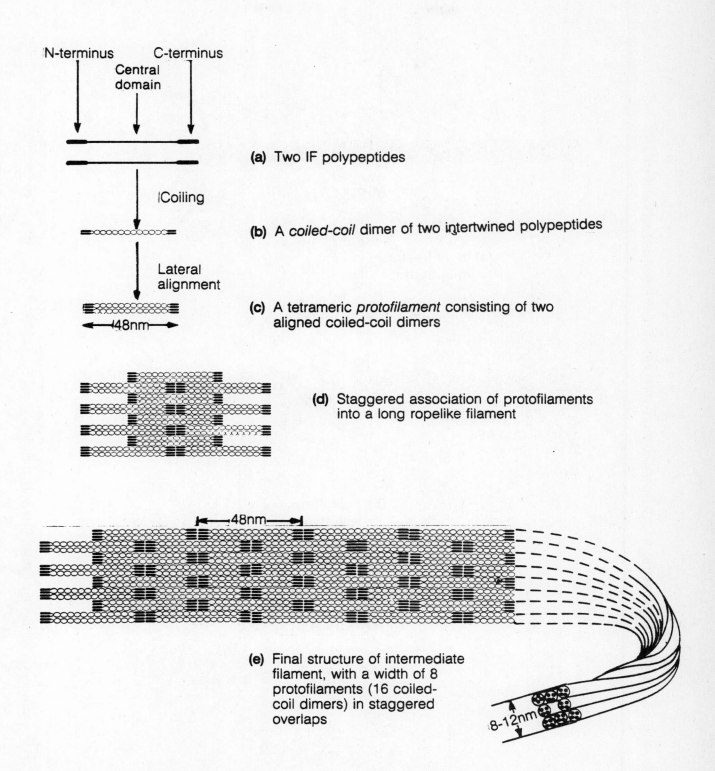

N-terminus   C-terminus
Central
domain

(a) Two IF polypeptides

Coiling

(b) A *coiled-coil* dimer of two intertwined polypeptides

Lateral
alignment

(c) A tetrameric *protofilament* consisting of two
aligned coiled-coil dimers

48nm

(d) Staggered association of protofilaments
into a long ropelike filament

48nm

(e) Final structure of intermediate
filament, with a width of 8
protofilaments (16 coiled-
coil dimers) in staggered
overlaps

8-12nm

Copyright © The Benjamin/Cummings Publishing Company, Inc.

Molecular Structure of Myosin

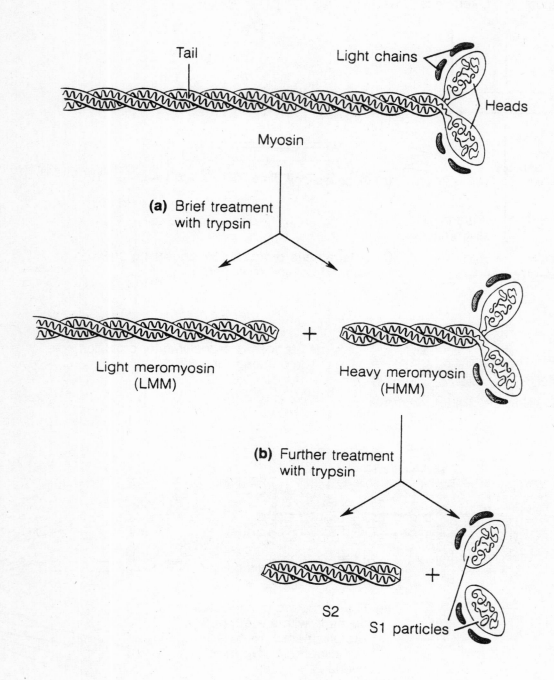

Copyright © The Benjamin/Cummings Publishing Company, Inc.

Levels of Organization of Skeletal Muscle Tissue

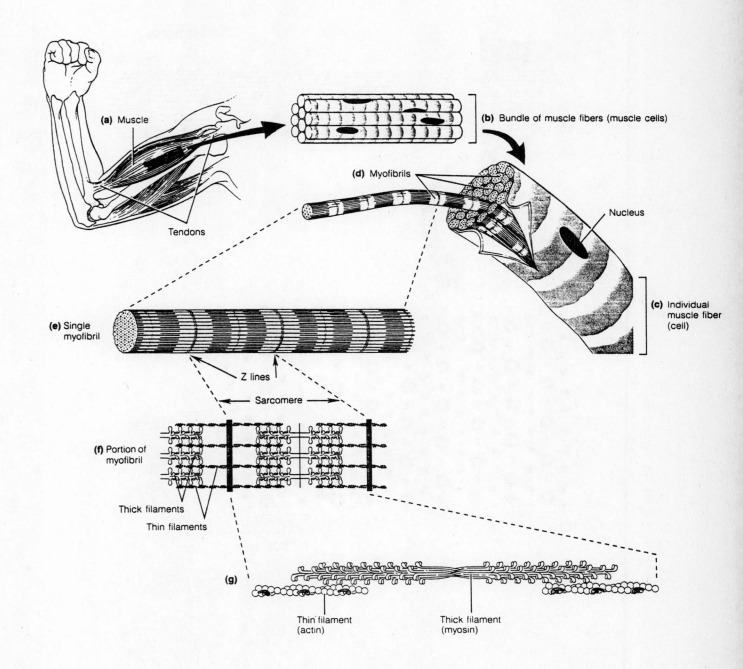

(a) Muscle

Tendons

(b) Bundle of muscle fibers (muscle cells)

(d) Myofibrils

Nucleus

(c) Individual muscle fiber (cell)

(e) Single myofibril

Z lines

Sarcomere

(f) Portion of myofibril

Thick filaments

Thin filaments

(g)

Thin filament (actin)

Thick filament (myosin)

Copyright © The Benjamin/Cummings Publishing Company, Inc.

Arrangement of Thick and Thin Filaments in the Myofibril

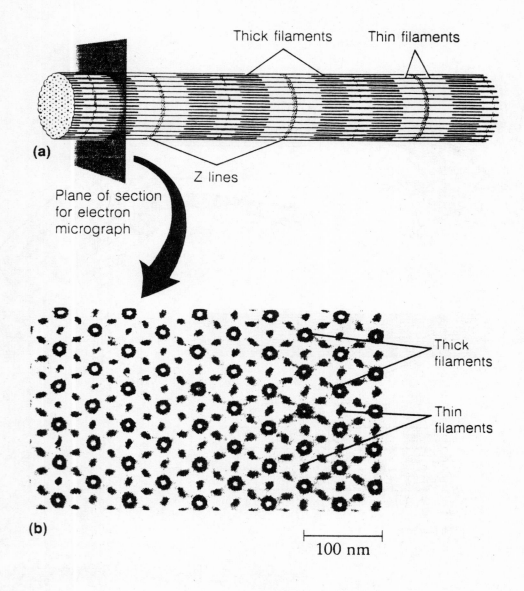

Thick filaments

Thin filaments

(a)

Z lines

Plane of section
for electron
micrograph

Thick
filaments

Thin
filaments

(b)

100 nm

Copyright © The Benjamin/Cummings Publishing Company, Inc.

Organization of the Thick Filament of Skeletal Muscle

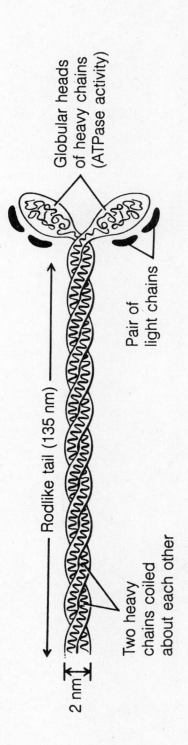

Globular heads
of heavy chains
(ATPase activity)

Pair of
light chains

Rodlike tail (135 nm)

Two heavy
chains coiled
about each other

2 nm

(a) The myosin molecule

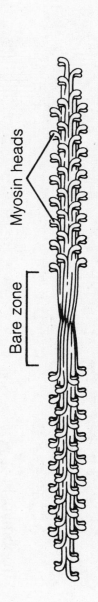

Myosin heads

Bare zone

(b) Organization of myosin molecules
into the thick filament of muscle

Copyright © The Benjamin/Cummings Publishing Company, Inc.

The Thin Filament of Striated Muscle

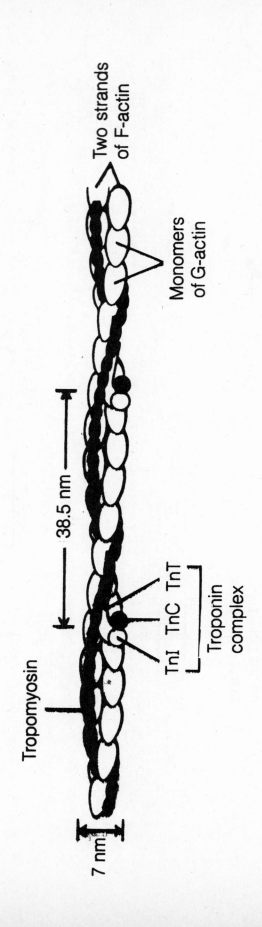

Two strands
of F-actin

Monomers
of G-actin

38.5 nm

TnI TnC TnT

Troponin
complex

Tropomyosin

7 nm

Copyright © The Benjamin/Cummings Publishing Company, Inc.

The Sliding Filament Model of Muscle Contraction

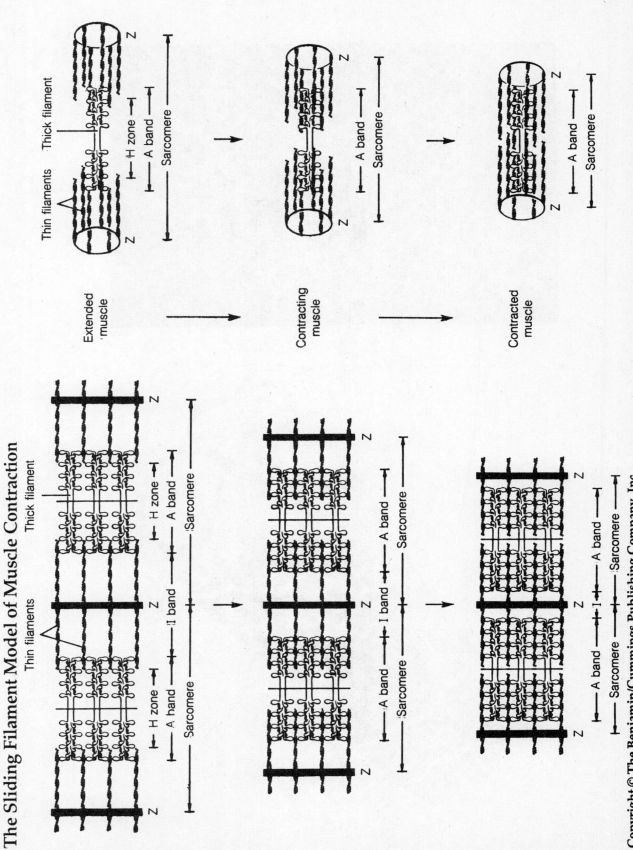

Copyright © The Benjamin/Cummings Publishing Company, Inc.

The Sarcoplasmic Reticulum and the Transverse Tubule System of Skeletal Muscle Cells

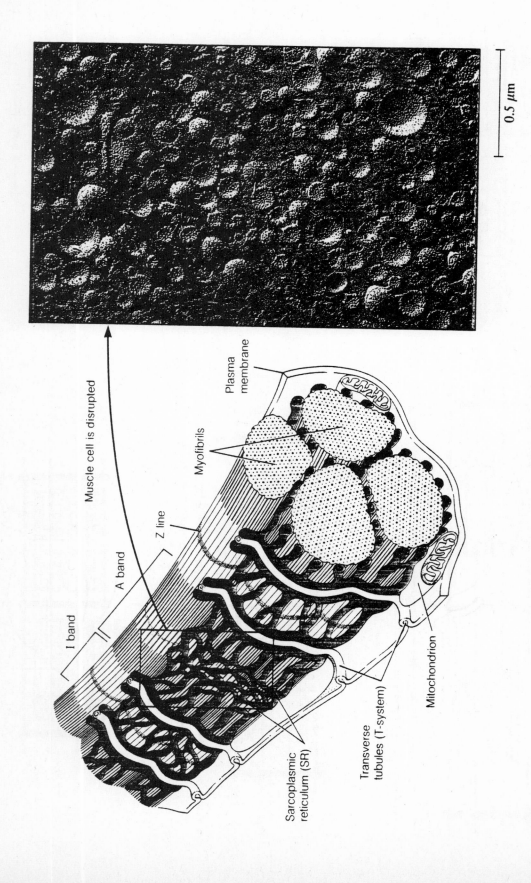

0.5 μm

Muscle cell is disrupted

I band

A band

Z line

Plasma membrane

Myofibrils

Sarcoplasmic reticulum (SR)

Transverse tubules (T-system)

Mitochondrion

Copyright © The Benjamin/Cummings Publishing Company, Inc.

Structure of the Calcium Pump Protein

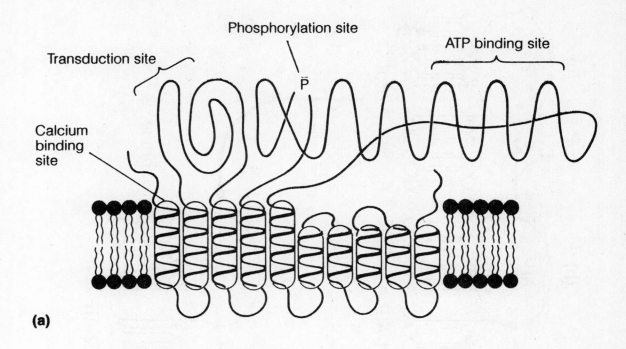

Transduction site

Phosphorylation site

ATP binding site

P̄

Calcium binding site

(a)

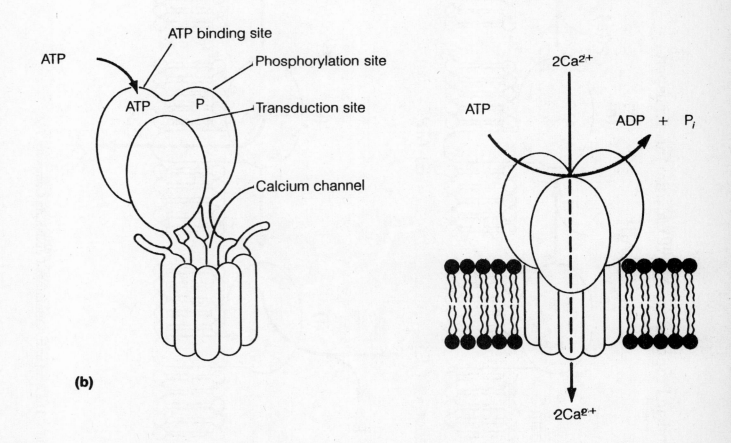

ATP

ATP binding site

Phosphorylation site

ATP

P

Transduction site

Calcium channel

(b)

ATP

$2Ca^{2+}$

ADP + P_i

$2Ca^{2+}$

Copyright © The Benjamin/Cummings Publishing Company, Inc.

Mechanism Proposed for Active Calcium Ion Transport

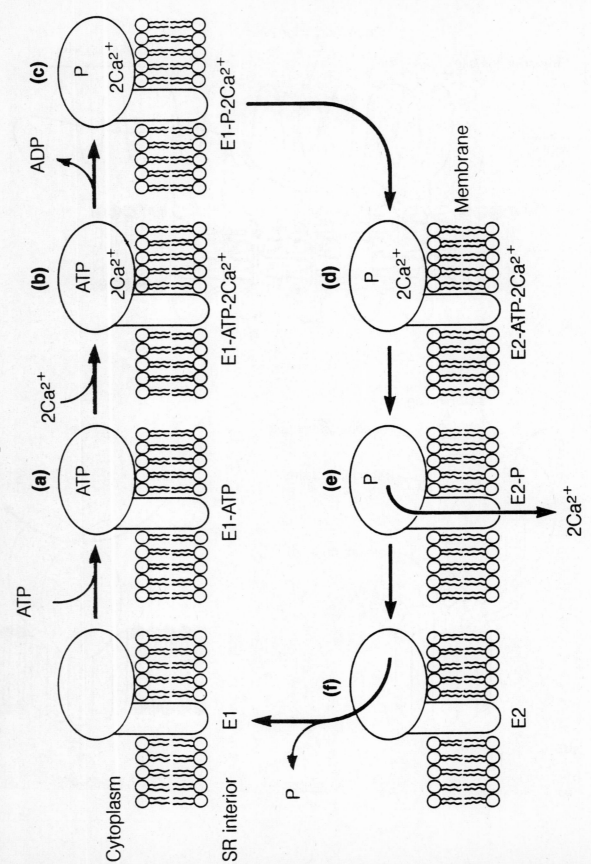

Copyright © The Benjamin/Cummings Publishing Company, Inc.

A Ciliated Protozoan

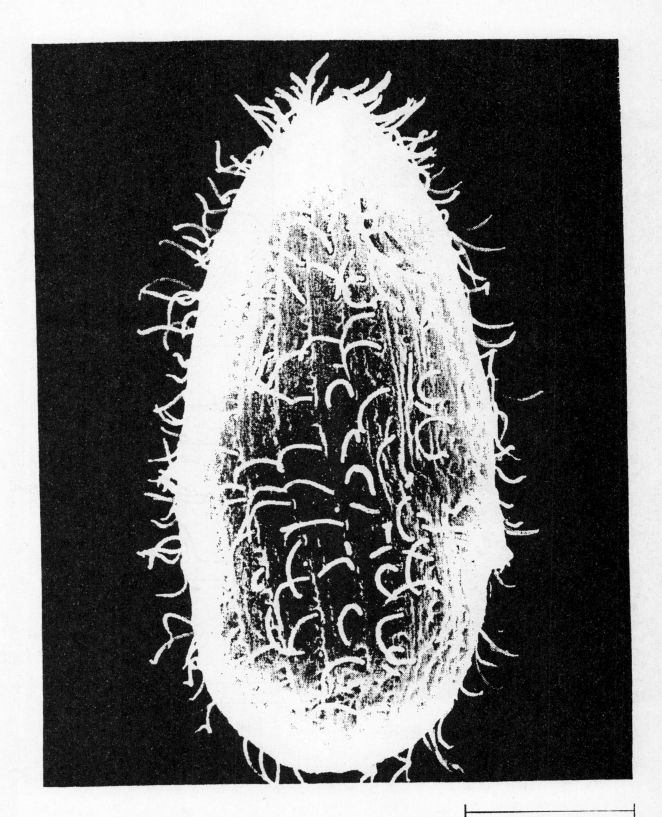

10 μm

Copyright © The Benjamin/Cummings Publishing Company, Inc.

The Beating of Flagella and Cilia

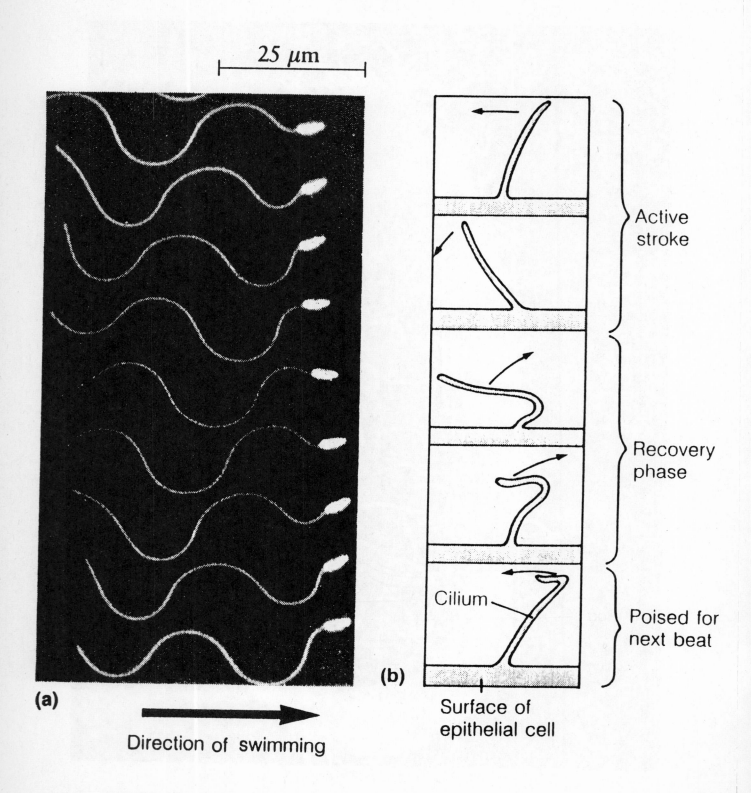

25 μm

(a)

Direction of swimming

(b)

Active stroke

Recovery phase

Cilium

Poised for next beat

Surface of epithelial cell

Copyright © The Benjamin/Cummings Publishing Company, Inc.

Structure of a Bacterial Flagellum and the "Motor" Responsible for Its Rotation

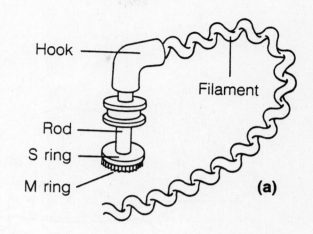

(a)

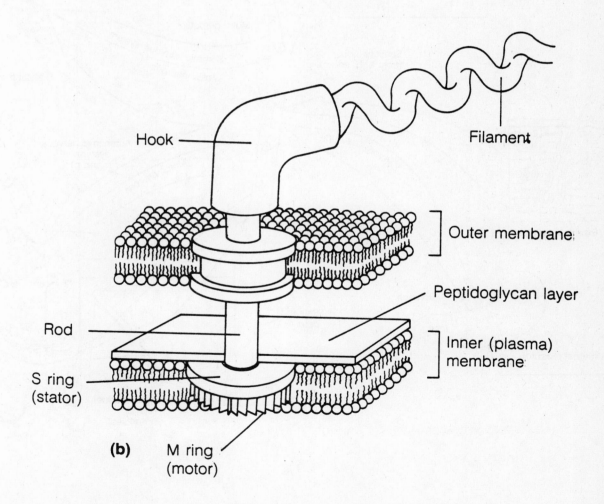

(b)

Copyright © The Benjamin/Cummings Publishing Company, Inc.

The Vertebrate Nervous System

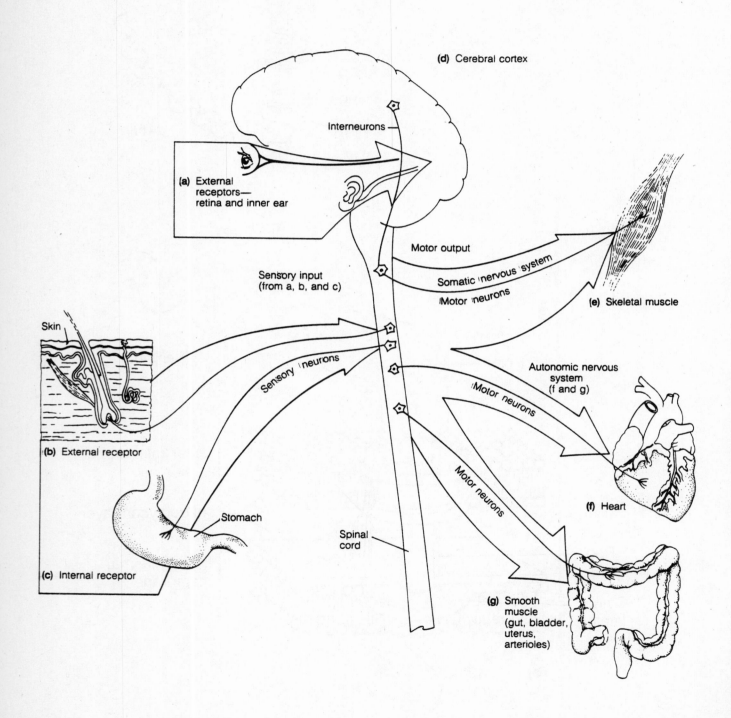

(d) Cerebral cortex

Interneurons

(a) External receptors— retina and inner ear

Motor output

Somatic nervous system

Motor neurons

(e) Skeletal muscle

Sensory input (from a, b, and c)

Skin

Sensory neurons

(b) External receptor

Autonomic nervous system (f and g)

Motor neurons

Stomach

(c) Internal receptor

Spinal cord

Motor neurons

(f) Heart

(g) Smooth muscle (gut, bladder, uterus, arterioles)

Copyright © The Benjamin/Cummings Publishing Company, Inc.

Structure of a Typical Motor Neuron

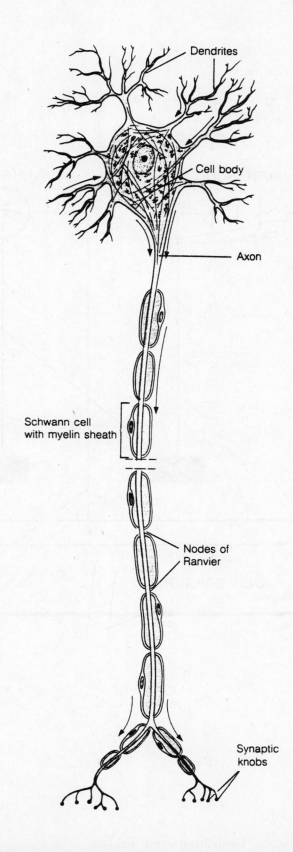

Dendrites

Cell body

Axon

Schwann cell
with myelin sheath

Nodes of
Ranvier

Synaptic
knobs

Copyright © The Benjamin/Cummings Publishing Company, Inc.

An Apparatus for Initiating and Measuring Changes in Membrane Potential

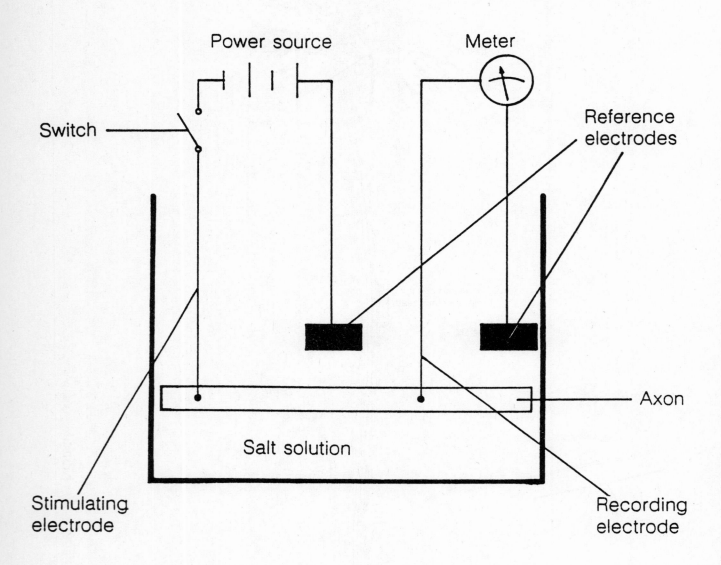

Copyright © The Benjamin/Cummings Publishing Company, Inc.

The Chemical Synapse

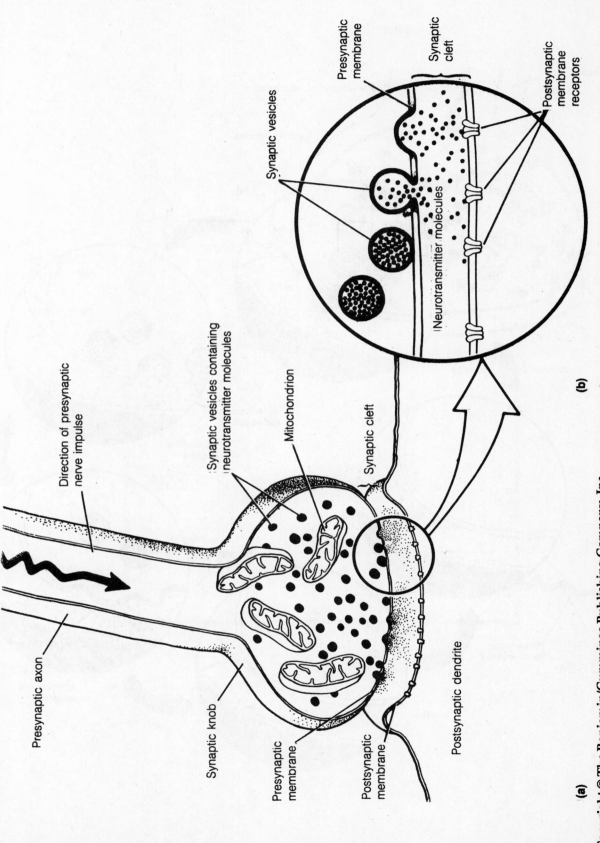

Presynaptic axon

Direction of presynaptic nerve impulse

Synaptic vesicles containing neurotransmitter molecules

Mitochondrion

Synaptic cleft

Synaptic knob

Presynaptic membrane

Postsynaptic membrane

Postsynaptic dendrite

(a)

Presynaptic membrane

Synaptic cleft

Synaptic vesicles

Neurotransmitter molecules

Postsynaptic membrane receptors

(b)

Copyright © The Benjamin/Cummings Publishing Company, Inc.

Transmission of a Signal Across a Cholinergic Synapse

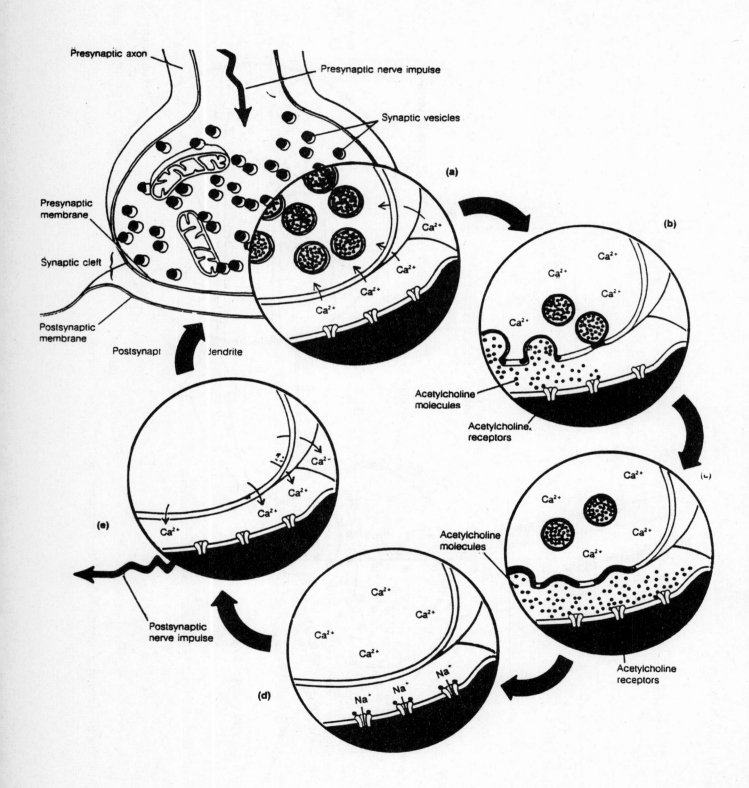

Presynaptic axon

Presynaptic nerve impulse

Synaptic vesicles

(a)

(b)

Presynaptic membrane

Synaptic cleft

Postsynaptic membrane

Postsynaptic dendrite

Acetylcholine molecules

Acetylcholine receptors

Acetylcholine molecules

Acetylcholine receptors

(c)

(e)

Postsynaptic nerve impulse

(d)

Copyright © The Benjamin/Cummings Publishing Company, Inc.

The Major Endocrine Tissues of the Human Body

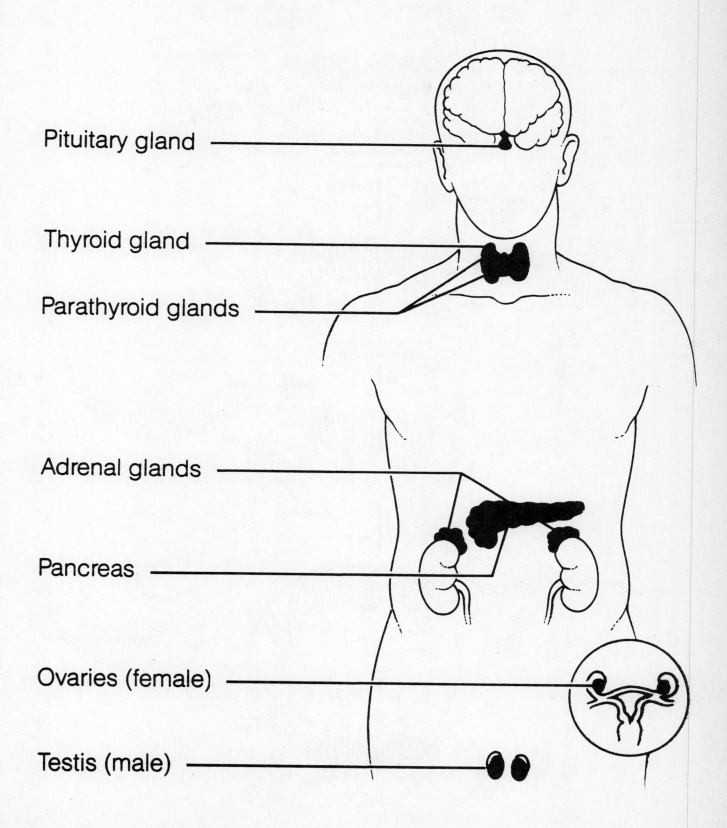

Pituitary gland

Thyroid gland

Parathyroid glands

Adrenal glands

Pancreas

Ovaries (female)

Testis (male)

Copyright © The Benjamin/Cummings Publishing Company, Inc.

The Role of G Proteins and Cyclic AMP in Signal Transduction

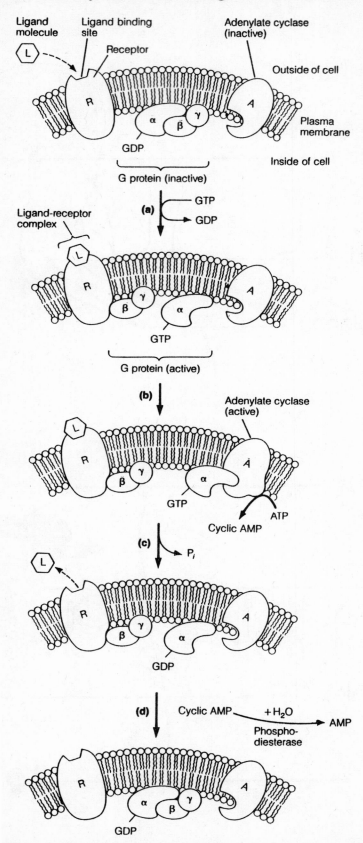

Copyright © The Benjamin/Cummings Publishing Company, Inc.

Stimulation of Glycogen Breakdown by Epinephrine

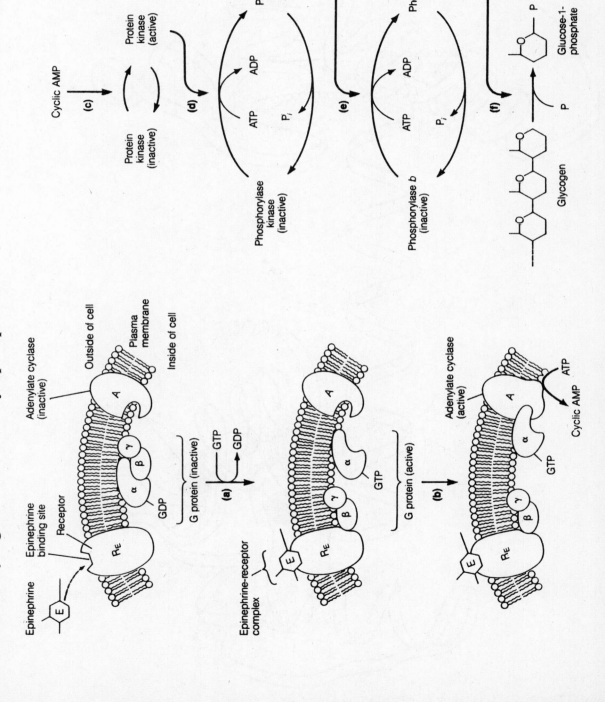

Copyright © The Benjamin/Cummings Publishing Company, Inc.

Structure of Calmodulin

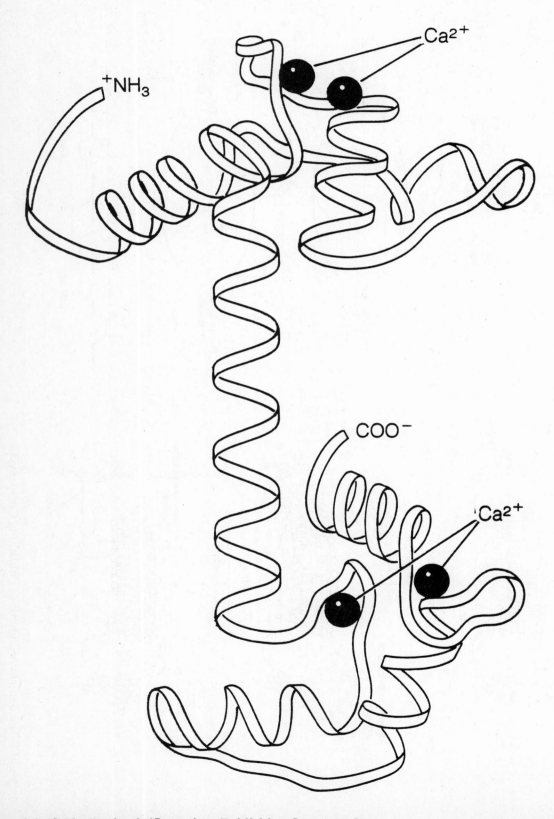

Copyright © The Benjamin/Cummings Publishing Company, Inc.

The Role of Inositol Trisphosphate and Diacylglycerol in Signal Transduction

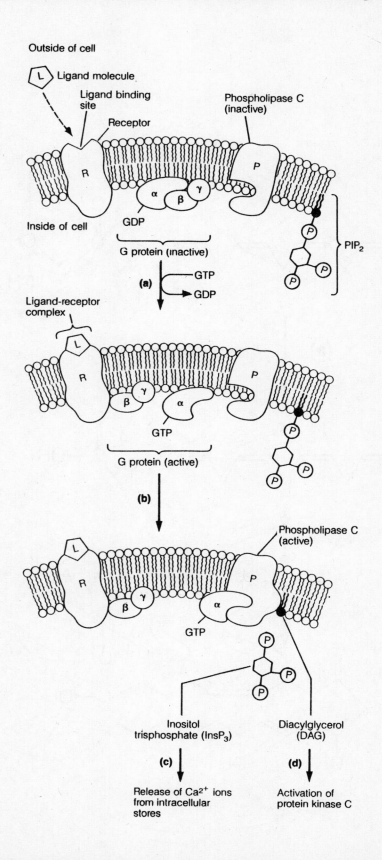

Copyright © The Benjamin/Cummings Publishing Company, Inc.

Phospholipase A and the Synthesis of Prostaglandins

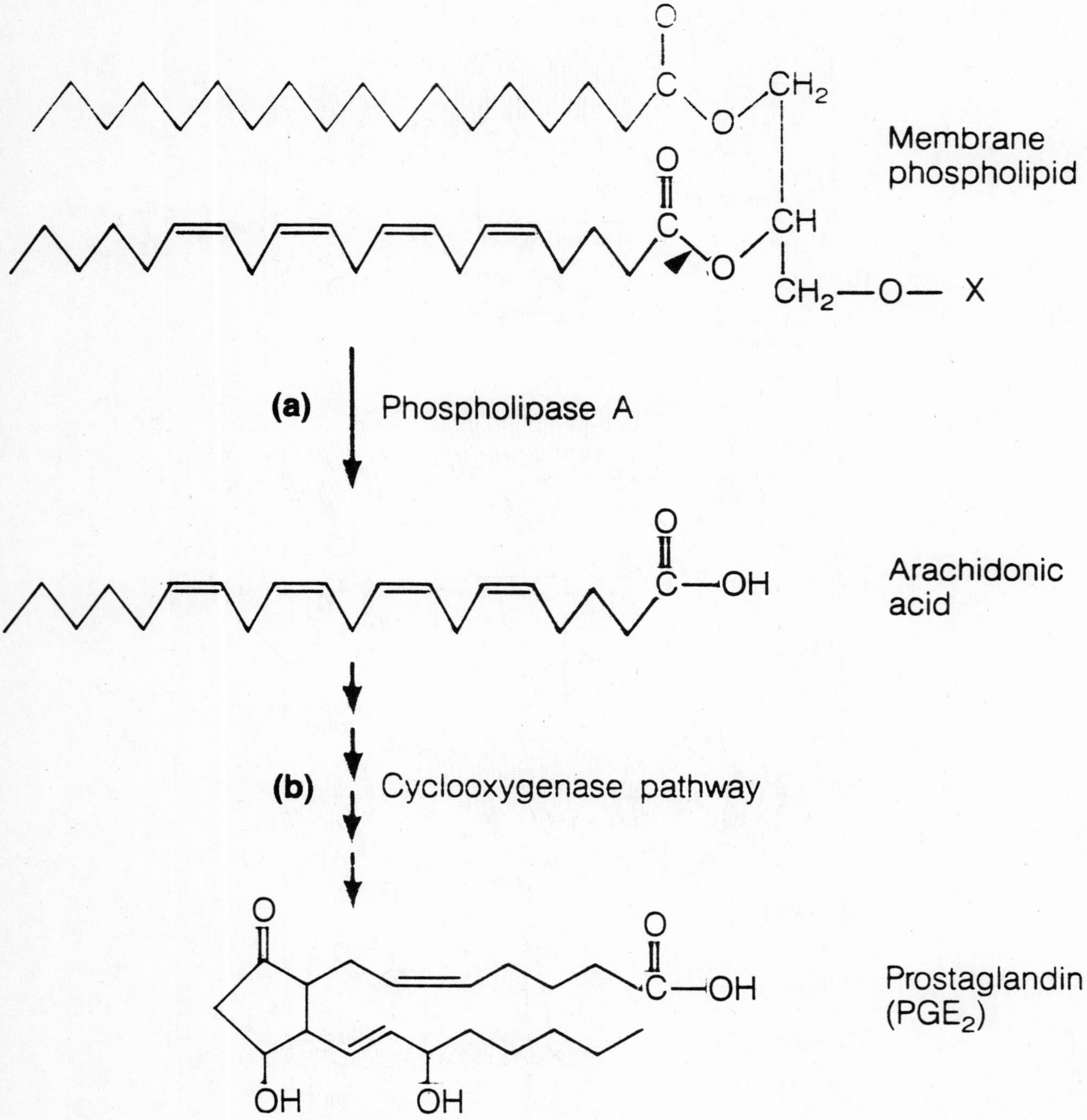

Membrane phospholipid

(a) Phospholipase A

Arachidonic acid

(b) Cyclooxygenase pathway

Prostaglandin (PGE$_2$)

Copyright © The Benjamin/Cummings Publishing Company, Inc.

Activation of Steroid Hormone Receptors

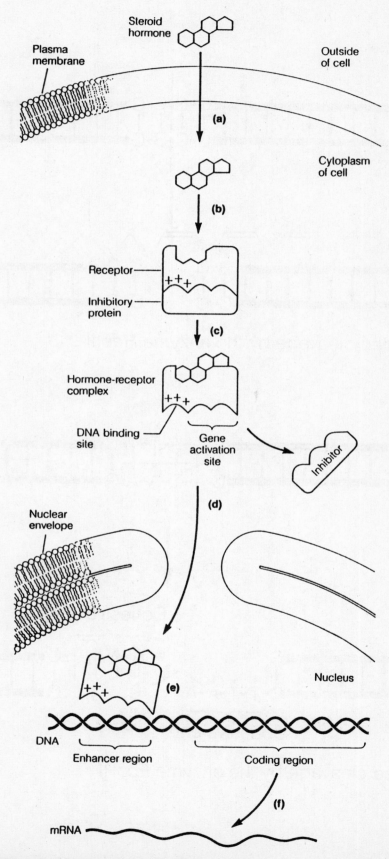

Copyright © The Benjamin/Cummings Publishing Company, Inc.

Cleavage of DNA by Restriction Enzymes

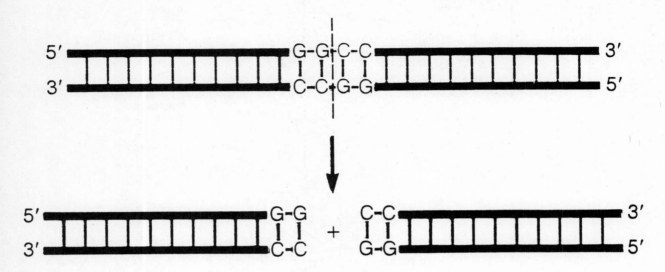

(a) Blunt-ended cleavage by the enzyme *Hae*III

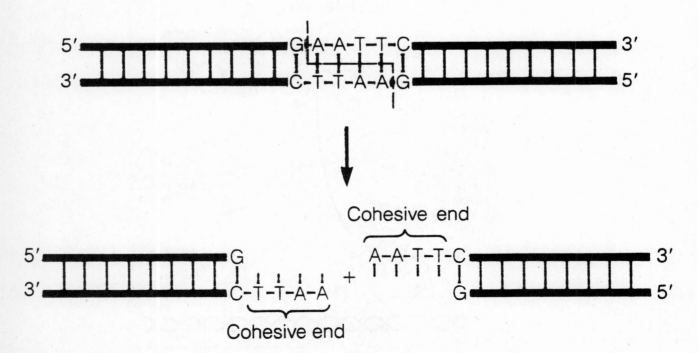

(b) Staggered cleavage by the enzyme *Eco*RI

Copyright © The Benjamin/Cummings Publishing Company, Inc.

DNA Cloning in Bacteria Using a Plasmid Vector

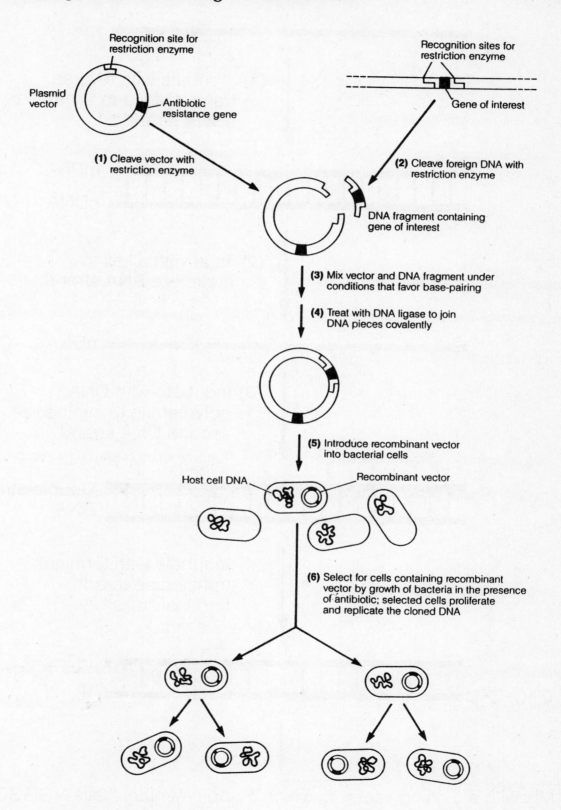

Copyright © The Benjamin/Cummings Publishing Company, Inc.

Preparation of Complementary DNA for Cloning

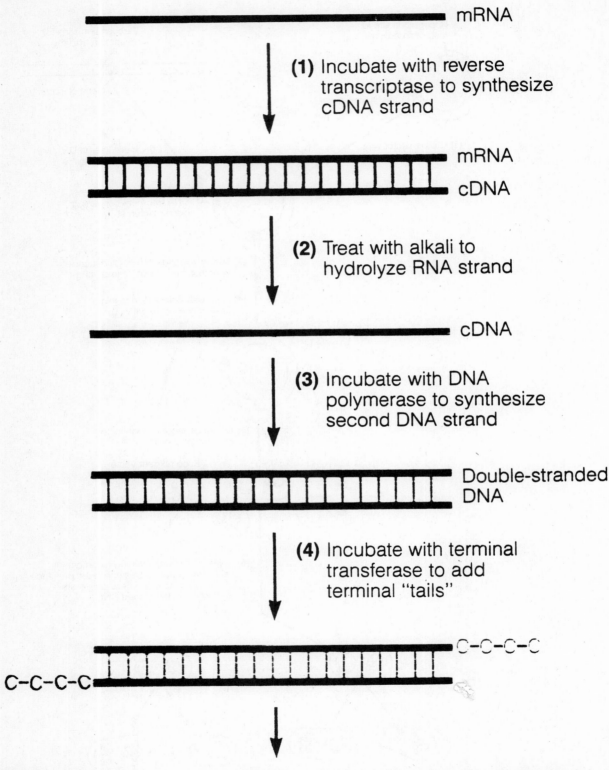

(1) Incubate with reverse transcriptase to synthesize cDNA strand

mRNA

mRNA

cDNA

(2) Treat with alkali to hydrolyze RNA strand

cDNA

(3) Incubate with DNA polymerase to synthesize second DNA strand

Double-stranded DNA

(4) Incubate with terminal transferase to add terminal "tails"

C–C–C–C

C–C–C–C

(5) Mix with a cloning vector to which complementary "tails" (GGGG) have been added and introduce recombinant cloning vector into bacterial cells as in figure B-7.

Copyright © The Benjamin/Cummings Publishing Company, Inc.